ELECTRICAL DESIGN,
SAFETY, AND
ENERGY CONSERVATION

**OTHER "JOB SIMULATION EXPERIENCE"
BOOKS BY THE AUTHOR**

Secrets of Noise Control
co-author Richard K. Miller, 1976
Fairmont Press, Atlanta, Ga.

*Plant Engineers and Managers
Guide to Energy Conservation,* 1977
Van Nostrand Reinhold, New York, N.Y.

Biorhythms and Industrial Safety, 1977
Fairmont Press, Atlanta, Ga.

How to Patent without a Lawyer, 1978
Fairmont Press, Atlanta, Ga.

ELECTRICAL DESIGN,

SAFETY, AND

ENERGY CONSERVATION

ALBERT THUMANN, P.E.

THE FAIRMONT PRESS

134 PEACHTREE STREET N. E. • ATLANTA, GEORGIA 30303

Contents

1. Reducing OSHA Related Electrical Violations   1

2. Using the Language of the Electrical Engineer   65

3. Applying the Fundamentals of Power   88

4. Analyzing Power Distribution Systems   109

5. How to Design a Lighting System   130

6. Conduit and Conductor Sizing   152

7. Using Logic to Simplify Control Systems   168

8. Applying Programmable Controllers and
   Electronic Instrumentation   192

9. Utilizing Intrinsic Safety in Hazardous Areas   204

10. Protective Relaying for Power Distribution Systems   212

11. Reducing Electrical Utility Costs   239

      References   249

      Index   251

Foreword

The understanding of electrical system design has become increasingly important not only to the electrical designer, but to safety, plant and project engineers as well. With the advent of the Occupational Safety and Health Act (OSHA), electrical citations have accounted for the single most common safety offender. Another factor which has caused plant and project engineers to become more aware of electrical systems has been high energy costs. Thus, to operate a plant efficiently, emphasis must be placed on energy conservation measures which affect electrical systems. Both safety and energy efficiency will be covered in this text along with the practical application problems for an industrial electrical design.

1

Reducing OSHA Related Electrical Violations

The Occupational Safety and Health Act (OSHA) has incorporated the National Electrical Code (NEC) into its standards. These OSHA standards were published in the 1974 *Federal Register* and are periodically updated. Thus a violation of the NEC is a violation of OSHA. In 1975 violations of the National Electrical Code were the most common violations of OSHA. Electrical violations were also number one in terms of penalties levied, with total assessments of almost $500,000.

As of the writing of this text, OSHA has adopted only the 1971 NEC specifications, even though 1978 NEC specifications are presently being used in industry. This has caused considerable confusion in industry. There is also confusion as to the degree of upgrading required as a result of modifications to an existing facility.

The material presented in this chapter has been adapted with permission from *Maintenance Engineering*, Cleworth Publishing Co., Inc., Cos Cob, Connecticut.[1,2,3,4,5]

1

OSHA: – RECOGNIZE AND CORRECT ELECTRICAL VIOLATIONS

Whether a plant's electrical system is only four years old and built in compliance with the 1968 NEC or if it is 35 years old and built in compliance with the 1937 NEC standards, certain areas must be up-dated to conform with the 1971 NEC specifications.

• Grounding of circuits, systems, tools, and portable equipment is now mandatory.

• "Temporary" hookups and installations must be eliminated as soon as possible. Make them by-the-book.

• Live parts of tools and electrical equipment must be properly guarded so they present no danger.

• Identify electrical equipment as to voltage rating and purpose. This not only warns unauthorized personnel to beware, but it also simplifies emergency repairs.

• Use flexible cords only for specific, temporary purposes. Never run them under work bins, across aisles or through any other place where they may likely get shorted out.

• Make sure electrical equipment used in a hazardous area is designed specifically for that purpose. Check especially the wiring, grounding, guarding, proper receptacles and plugs, drainage, and sealing methods.

• Modify any installation that might put an employee into danger.

OSHA's final interpretation of what is dangerous and in violation and what is acceptable depends greatly upon the individual inspector who does a "walk-around" in the plant. Even if the equipment and wiring are UL-listed, he may not like a particular installation and issue a citation anyway. Since there is a disagreement on some of the interpretations of the NEC, it might be valuable to subscribe to the *B. N. A. Reporter.* Published by the Bureau of National Affairs, Washington, D.C., this loose-leaf publication gives a weekly summary of any event (an appeal of a citation, change in interpretation by OSHA, lists of fines, etc.). This keeps a plant engineer or safety director up-to-date on the law and can perhaps save a fine.

Table 1-1. Most Commonly Cited Electrical Violations

A typical plant has many violations that can be cured by better housekeeping procedures. When an area is cleaned-up, a sloppy electrical installation tends to "jump out at you." Although the fines that can develop are often costly, the problems themselves usually take little time to abate.

Water or oil on floor around electrical machinery. Wipe it up and fix the leak.

Exposed wiring must be labelled as to where it goes and how much power. Replace covers on junction boxes and electrical enclosures.

Unlocked door and lack of a warning sign. The service entrance door must be locked and labelled to protect unauthorized personnel. Clearly mark each piece of electrical equipment as to its use.

Empty sockets should be permanently removed if they are not used for lighting. If they are used for lighting a bulb should be inserted.

Ungrounded electrical tools must have a third-wire ground. (If the tool is a self-grounding type, it must be clearly identified.) Make sure your grounding plugs haven't been deliberately sawed or broken off for convenience. If missing a third prong, replace the plug.

Extension cords must be removed from under doors, across aisles, inside storage areas, or wherever movement of materials (or wheels of a fork lift truck) can damage the cord. Replace with permanent wiring or cord set designed for aisle use.

Loose connections should be tightened. This insures a low resistance impedance which facilitates the operation of overcurrent devices. Check connections with an ohmmeter and make connections "wrench-tight."

Worn or frayed cables anywhere in plant should be replaced.

Live metal lamp guards or live cases for lamps should be repaired at once. Defect is in the insulation, broken plug, or shorted receptacle.

Unlabelled 220 volt outlets can be mistaken for a normal 110 volt unit. Identify them so no confusion will ever exist.

Bridged fuses must be cleared.

Portable, hand-held lamps must be of molded composition—not the old paper-lined or brass-shell compositions. Don't use the old units.

Attachment plugs must be constructed for rough use and equipped with a cord grip that does not put strain on the terminal screw.

Work habits. Educate employees to be electrically-safe in their work habits. Be sure the equipment they are using is safe, and that they wear proper safety gear.

DEGREE OF COMPLIANCE CAN VARY

An in-plant inspection should be done before an OSHA inspection is made.

When an electrical inspector is appointed by the plant engineer to check out the facility for possible violations, it must be kept in mind that OSHA does not require the same degree of compliance with every type of electrical installation. The inspector must realize there is no intent by OSHA to force a plant to rip out any old installations just to up-date them. The purpose is electrical safety. Of course, all systems must comply with the previously mentioned retroactive regulations.

• Existing installations that are in-place before March 15, 1972 can use replacement devices of the same type being presently used, unless specifically prohibited by the code. Extending branch circuits, shifting equipment within a plant, and replacing wiring plugs and receptacles are minor modifications that are included under this category. Some of the replacement devices may not even carry a UL-label anymore, but are still permitted in this application.

• Major modifications, repairs, and rehabilitations such as tearing down a wall, adding a new wing, or running a new power supply into a plant must follow the 1971 NEC.

• Job site wiring, both temporary and permanent, must conform to pertinent sections of the new NEC. A federal inspector will most likely put emphasis on the grounding requirements.

• New electrical work, of course, follows the latest code.

Keeping the above requirements in mind as an electrical inspection is made, may make it possible to save unneeded purchases of equipment by properly defining what kind of installation is being dealt with. But also remember, any installation that is in any way hazardous must be changed. If in doubt, bring it up to the newest NEC standards.

USE THE BEST ELECTRICAL INSPECTOR

Only someone thoroughly familiar with the NEC and capable of using electrical test equipment can hope to do a proper check for electrical compliance. But, even so, years of living with the same electrical violations can sometimes go unnoticed by the men who work around and on them. A large maintenance department has the luxury of bringing a technician in from another area; a small department must encourage their inspector to examine their facility in a "fresh light."

This "new-man" approach is similar to a real OSHA inspection, and it can lead to some startling discoveries. Maybe the storage of fuel or chemicals into a new area has changed the Hazardous Location classification from a Class II to a Class I. Or maybe the relocation of the welding crew's equipment has changed the classification of an area. This is the sort of thing more easily picked up by this fresh approach. A check must be made on all fittings, wiring methods, enclosures, and equipment to be sure that they comply.

The scope of the electrical check-out takes the greater part of an inspection. Personnel interviewed at one plant spent about 300 hours just interpreting the rules and establishing electrical check lists. In addition, local and state codes must be checked to be sure you are meeting their requirements. (Comply with OSHA if they are different in any way.)

DECIDE IF THE INSTALLATION IS ACCEPTABLE

"Approved" means acceptable to the authority enforcing the code, the representative of the Department of Labor. It meets his requirements only if:

• It has been certified, listed, labelled, or otherwise determined to be safe by Underwriters' Laboratories, Inc., Factory Mutual Engineering Corporation, or another nationally recognized testing body.

This means the purchasing department can prevent compliance headaches later on by specifying tested products from the start.

• Equipment that is not checked out by a nationally-recognized testing laboratory is determined to be safe if it has been inspected by another federal agency, or by the state, local, or municipal authority responsible for enforcing the safety provisions of the NEC.

All high voltage switchgear, and most low voltage switch-gear, switchboards, and motor control equipment is not in-spected by UL or Factory Mutual.

• Custom-made equipment is determined to be .safe for its intended use on the basis of the manufacturer's test data. The employer must keep this information available for inspec-tion by any representative of the Department of Labor.

The company's design engineer has ultimate responsibility for OSHA compliance of the equipment he specifies. All high voltage switchgear, most low voltage switchgear, and related equipment are assembled for a particular job, placing them in the "custom-made" category of OSHA. Some low voltage sec-tions can be UL-listed for specific requirements, but if a non-listed part is contained within the system, the label is generally withheld. The data that is needed from the manufacturer relates to construction, tests and safety provisions of this custom equipment.

COMPILE ELECTRICAL CHECK LISTS

Once it has been determined that the general physical lay-out of the plant is safe from the most common (and easily abated violations) it is necessary to check out NEC compliance (Table 1-2).

Make sure all electrical equipment is properly mounted—firmly with no wooden plugs driven into holes in masonry, con-crete or similar materials. The mechanical execution of all work must be neat and by-the-book. Any temporary wiring situation that has lasted years must be eliminated. Answer check list ques-tions either yes (out of compliance) or no (in compliance). If a doubt exists in your mind, use the more specific check lists for overcurrent devices, transformers, grounding, wiring, etc. If a doubt still exists, mark it out-of-compliance and check it out later.

The electrical check list that is developed for the plant should be a personalized tool. The value of the list is increased if the plant's foreman and technicians review them from time-to-time (the same as they do for a preventive maintenance job) to make sure the plant remains in compliance.

Table 1-2. OSHA Check List—National Electrical Code

(Write Yes or No, Describe Equipment, Location)

• Are there installations and/or equipment in the plant that are not installed and maintained in accordance with the latest N.E.C., N.F.P.A., and ANSI?

CONDUCTORS

• Are there conductors used in the plant which are of materials other than copper?

DETERIORATING AGENCIES

• Are there conductors or equipment used in damp or wet locations, exposed to agents having a deteriorating effect on them, which are not approved for use in these areas?

MECHANICAL EXECUTION OF WORK

• Are there installations and/or equipment which is not installed in a neat and workmanlike manner?

MOUNTING OF EQUIPMENT

• Is there any equipment not firmly secured to the surface on which it is mounted?

CONNECTIONS TO TERMINALS

• Are there any connections of conductors to terminals which have been made by other means than pressure connector (including set screw type), solder, lugs, or splices to flexible leads?

(EXCEPTION: No. 8 or smaller solid conductors and No. 10 or smaller stranded conductors may be con-

nected by means of clamps or screws with terminal plates having upturned legs.)

• Are there terminals being used for more than one conductor which is not of a type approved for this purpose?

• Are there installations and/or equipment where terminating devices such as pressure connectors and solder lugs have been used which are not suitable for the material and/or the size of the conductor?

• Are there installations and/or equipment where solder flukes, inhibitors or other compounds have been used that adversely affected the conductors, installation, or equipment?

SPLICES

• Are there conductors which have been spliced or joined by means other than with splicing devices approved for the use by brazing, welding, or soldering with a fusible metal or alloy?

• Are there soldered splices which have not been joined such that they were mechanically and electrically secure prior to the application of the solder?

Are all splices and joints and free ends of conductors which are not covered with an insulation equivalent to that of the conductors?

WORKING SPACE AROUND ELECTRICAL EQUIPMENT (600 VOLTS OR LESS)

• Are there installations and/or equipment where sufficient access and working space is not provided and

/more/

Table 1-2. OSHA Check List—National Electrical Code (continued)

maintained to permit safe operation and maintenance?

• Are there installations and/or equipment which does not comply with the following requirements:

WORKING CLEARANCE

Except as elsewhere required or permitted in this sub-part, the dimension of the working space in the direction of access to live parts operating at no more than 600 volts, which are likely to require examination, adjustment, servicing or maintenance while alive, shall not be less than indicated in the following table. Distances are to be measured from the live parts if such are exposed or from the enclosure front or opening when such are enclosed.

VOLTAGE TO GROUND	MINIMUM CLEAR DISTANCE CONDITION (ft)		
	1	2	3
0 - 150 volts	2½	2½	3
151 - 600 volts	2½	3½	4

• Condition 1 is an exposed live part on one side and no live or grounded part on the other side of the working space, or exposed live parts on both sides effectively guarded by suitable wood or other insulated wire or insulated bus bars (operating at not more than 300 volts) are not be considered live parts.

them, which do not have a minimum of 3 feet of front working space?

ILLUMINATION

• Are there working spaces about switchboards and control centers which do not have adequate illumination?

HEADROOM

• Are there working spaces about switchboards and control centers where there are live parts exposed at any time which do not have a minimum of 6¼ feet of headroom?

GUARDING OF LIVE PARTS (NOT MORE THAN 600 VOLTS)

ENCLOSURES

• Are there installations and/or equipment operating at more than 50 volts which are not guarded against accidental contact by approved enclosures or any of the following methods:

    (a) By location in a room, vault, or similar enclosure which is accessible only to qualified personnel.

    (b) By suitable permanent, substantial partitions or screens so arranged that only qualified persons will have access to the space within reach of the live parts. Any openings in such partitions or screens shall be so sized and located that persons are not likely to come into accidental contact with

/more/

Table 1-2. OSHA Check List—National Electrical Code (continued)

• Condition 2 is exposed live parts on one side and grounded parts on the other side. Concrete, brick, or tile walls are considered grounded.

• Condition 3 is exposed live parts on both sides of the work space (not guarded as in #1) with the operator between.

(Exception: Working space is not required in back of assemblies such as deadfront switchboards or control centers when there are no renewable parts such as fuses or switches on the back and when all connections are accessable from other locations than the back. Smaller spaces may be permitted where it is judged that the particular arrangement of the installation will provide adequate accessibility.)

CLEAR SPACES

• Are there working spaces as required being used for storage?

• Are there working spaces, where normally exposed live parts are exposed for inspection or servicing, which are located in passageway or general open spaces?

ACCESS AND ENTRANCE TO WORKING SPACE

• Are there working spaces about electrical equipment which do not have at least one entrance of sufficient area to allow working space?

FRONT WORKING SPACE

• Are there switchboards and control centers, where there are live parts normally exposed on the front of the live parts or to bring conducting objects into contact with them.

(c) By a guard rail, provided the live parts operate at 600 volts or less and provided the location is such that it makes contact with live parts unlikely.

(d) By location on a suitable balcony, gallery, or platform so elevated and arranged as to exclude unqualified persons.

(e) By elevation at least 8 feet above the floor or other working surface.

GUARDS

• Are there installations and/or equipment exposed to potential physical damage which are not protected by enclosures or guards so arranged and of such strength as to prevent such damage?

ENTRANCES

• Are there entrances to room and other guarded locations containing live exposed parts which are unmarked with conspicous warning sign forbidding unqualified persons to enter?

ARCING PARTS

• Are there parts of electrical equipment (excluding the welding equipment) which in ordinary operation produce arcs, sparks, flames, or molten metal which are not enclosed or separated and isolated from all combustible material?

|more|

Table 1-2. OSHA Check List—National Electrical Code (continued)

MARKING
• Is there equipment which is not marked with the manufacturer's name, trademark, or other descriptive marking by which the organization responsible for the product may be identified?
• Is there equipment which is not marked with the voltage, current, wattage, or other electrical ratings?
IDENTIFICATION
• Are there disconnecting means for motors and appliances and service, feeder, or branch circuit (at the point where they originate) which are not legibly marked to indicate its purpose, unless located and arranged so that the purpose is evident?
• Are there markings on equipment which are not sufficiently durable to withstand the environment involved?

OVERCURRENT PROTECTION
PROTECTION OF EQUIPMENT
• Is there any equipment not protected against overcurrent?
INTERRUPTING CAPACITY
• Are there devices intended to break current which do not have sufficient interrupting capacity for the voltage employed and for the current which must be interrupted?

LOCATION IN PREMISES
• Are there overcurrent devices located where they are exposed to physical damage.
• Are there overcurrent devices located in the vicinity of easily ignitible materials?
ENCLOSURES FOR OVERCURRENT DEVICES
• Are there overcurrent devices, other than those which are a part of a specially approved assembly which affords equivalent protection or unless mounted on switchboards, panelboards, or controllers located in rooms or enclosures free from easily ignitible or dampness, which are not enclosed in cutout boxes or cabinets?
(NOTE: The operating handle of a circuit breaker may be accessible without opening a door or cover.)
• Are there enclosures for overcurrent devices in damp or wet locations which are not of the type approved for such locations?
• Are there enclosures in damp or wet locations which are not mounted so there is at least ¼-inch air space between the enclosure and the wall or other supporting surface?
ARCING AND SUDDEN MOVING PARTS
• Are there fuses and/or circuit breakers which are not located and shielded such that persons will not be burned or otherwise injured by their operation?

/more/

Table 1-2. OSHA Check List—National Electrical Code (concluded)

CIRCUIT IMPEDANCE AND OTHER CHARACTERISTICS

• Are there circuit protective devices which will not clear a fault without the occurrence of extensive damage to the electrical components of the circuit? (NOTE: This fault may be assumed to be between two or more of the circuit conductors, or between any circuit conductor and the grounding conductor or enclosing metal raceway.)

SUDDEN MOVING PARTS

• Are there handles or levers of circuit breakers, and similar parts which move suddenly in such a way that a person in the vicinity is liable to be injured by being struck by them, which are not guarded or isolated?

Final acceptance of the electrical work in the plant is the prerogative of the electrical inspector from the Department of Labor. Particular site, construction, or installation problems in the plant may cause seemingly good electrical work to be rejected. Since the local OSHA office will issue the citations, it's a good idea to contact them whenever questions arise.

OSHA: – INSTALL AND MAINTAIN
VALID ELECTRICAL GROUNDS

Use the OSHA approach to determine if the plant has installed and is properly maintaining valid electrical grounds that comply with the 1971 *National Electrical Code.* To make a proper visual and physical inspection instruments are needed such as:

• A ground circuit tester that plugs into electrical outlets. A signal light indicates an open circuit, a shorted wire, an improper hook-up, or an incorrect polarity of wires.

• A bio-medical field probe that gives off a loud alarm signal when placed near any machine that gives off static or has current leakage. Pull out the antenna and carry it around the plant, placing it nearby potentially dangerous tools and machinery.

• A ground loop impedance tester that plugs into an electrical outlet. This not only determines if a ground exists, but it gives a direct resistance measurement.

The fault problem is either in the service (system) or the equipment ground.

Service or system grounds have one of the circuit power conductors grounded. This wire, designated the "identified" or "grounded" conductor, is color-coded white. Thus, in an ordinary low voltage circuit, the white wire is always grounded at the transformer and generator, and again at the building's service entrance. While this ground protects the machine or tool, a second wire must be provided from the metallic non-current carrying parts to ground for the protection of personnel.

This second wire transmits stray (fault) currents that may enter the metal parts safely away from the user. The resulting heavy surge of current operates the circuit protective devices and opens the circuit. The faster this takes place, the better the protection.

In the case of permanent or fixed equipment, a low resistance conductor is attached to the metallic non-current carrying part of the equipment, and then to a cold water pipe, or artificial ground. Make sure the conductor has a low resistance, offers a continuous path, and is installed where it would be unlikely to be damaged.

Grounding the machine is the usual method of giving protection to users of power tools. Because the operator moves about so much, a fixed grounding conductor is impractical. The ground wire offers personnel protection if it has a low enough resistance, since the unwanted current will take this path of least resistance to ground.

The resulting surge of current opens the overcurrent devices and de-energizes the circuit. Don't confuse this with circuit grounding which primarily protects the circuit and equipment rather than personnel.

Machine grounding is done in various ways:

(1) A polarized system has three wires contained within a flexible cable. One of these wires (green) is not part of the circuit but is placed in the cable for the sole purpose of grounding the tool in the event it gets energized. This wire is connected permanently to a non-current carrying metal part of the tool; at its plug end, it's connected to a prong which is separate from the prongs attached to the current carrying wires which constitute the circuits. Plug into a receptacle and you have connection with a permanently-grounded conductor.

(2) A metallic conduit circuit, without a third wire is used exclusively for equipment grounding. The three-wire flexible cord is still employed, but the preferred method is to use a special receptacle designed for the cord. The contact point into which the grounding prong of the flexible cord is inserted is directly connected to the metallic conduit or to the metallic

junction box. Since the metallic circuit is grounded, plugging in the receptacle completes the circuit.

(3) A metallic conduit circuit, without a special receptacle can establish an acceptable ground by using a grounding adapter in conjunction with a three-wire flexible cord and a three-prong plug. Such adapters can be:

• A plug with a built-in pigtail grounding connection that attaches to the receptacle by loosening the screw holding the faceplate to the outlet box, attaching pigtail to screw, and re-tightening it.

• A duplex grounding adapter that plugs into the wall receptacle and grounding contact is made through a long screw which replaces the one holding the faceplate to the receptacle.

Both these adapters establish their ground through the metallic conduit (which is usually grounded at the switch or fuse box). Adapters should remain attached to receptacles, so that it is not necessary to make a connection each time the tool is needed.

(4) A non-metallic cable is used, or a metallic cable cannot be grounded. Neither the three-wire plug with special receptacle, adapter plug, nor the two-wire plug with the extra ground wire can be effectively used in this case. The ground wire should be attached directly to a water pipe or to an artificial ground.

MAKE LOW RESISTANCE CONNECTIONS

The less the resistance is in a path to ground, the more current it can carry and the more effective the ground. This ground wire must be of sufficient size to provide low resistance (preferably not more than 3 ohms), and it should be installed in such a manner that it is protected from damage.

Grounding to a cold water pipe is ideal and should be employed where practical. Make sure the ground is attached to the street side of the valve, otherwise it is possible to lose the ground when water is turned off and the continuity is destroyed by pipe calking compounds.

In some cases a buried artificial ground may be necessary. This consists of a ground wire connected to an electrode driven

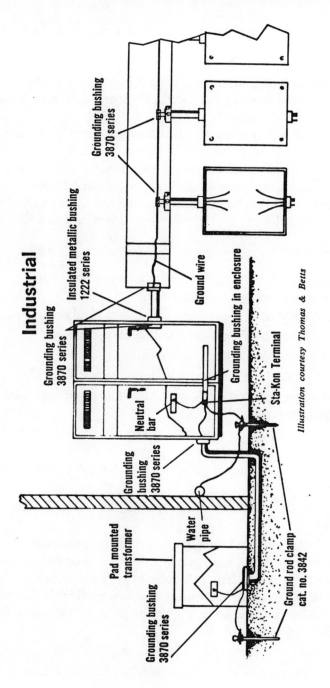

Figure 1-1.

at least 8 feet into the ground, or a metal grid buried in the earth. When an artificial ground is used, resistance should be preferably less than 25 ohms.

All ground connections should be made and checked by a person with knowledge of electrical theory. Check each connection with an ohmmeter.

In providing the ground, remember that the lower the resistance, the greater the current carrying capacity. This is important because if a person comes into contact with an energized tool or machine (even though grounded) and he has a low resistance, a portion of the current will flow through him instead of the ground wire. This portion may be enough to kill him.

WHAT MUST BE GROUNDED

Fixed electrical equipment must be grounded if:
- Supplied by iron-clad wiring.
- Located in a wet location and not insulated.
- In reach of a person standing on the ground.
- Located in a hazardous location.
- In electrical contact with metal or metal lath.
- Terminal voltage is over 150 volts to ground, with these exceptions:

(a) Enclosures for switches and circuit breakers accessible only to qualified personnel.

(b) Metal frames of electrically-heated devices insulated from ground.

(c) Transformers mounted higher than 8 feet above ground.

Frames of stationary motors must be grounded when supplied by metal-clad wiring; located in a wet or hazardous area; operated at more than 150 volts to ground; or not isolated or guarded. Grounding the motor frame is preferred. But in cases where this can't be done, it must be permanently and effectively insulated from ground.

Some non-electrical equipment must be grounded:
- Frame and track of electrically-operated cranes.

• Frame of non-electrically-driven elevator-car to which electrical conductors are attached.

• Hand-operated metal shifting ropes or cables of electrical elevators.

• Partitions, grill work and other enclosures around electrical equipment carrying in excess of 750 volts between conductors (unless in substations or vaults under the sole control of the supply company).

• Service and other conductor enclosures such as raceway, metal service cable sheathing, or armoring. In runs less than 25 feet which are free from probable contact with ground, grounded metal, metal lath, or conductive thermal insulation, and where personnel are guarded against accidental contact, grounding is not required. Keep the metal enclosures of conductors at least 6 feet away from lightning rod conductors. If this distance is too great, bond them together.

EQUIPMENT CONNECTED BY CORD AND PLUG

Exposed non-current carrying metal parts of cord and plug connected equipment must be grounded:

• If used in hazardous (including wet) areas.

• When operated at more than 150 volts to ground (except guarded motors and metal frames of electrically-heated appliances).

• If refrigerators, air conditioners, sump pumps, drills, sanders, wet scrubbers, and saws are used in other than a residential location.

• If portable tools are likely to be used in wet or conductive locations unless supplied through an insulating transformer with an ungrounded secondary of not over 50 volts (exceptions are noted in the NEC).

• Metal frames of portable, stationary and fixed electrically-heated appliances operating above 150 volts to ground circuits must be grounded in the manner specified by the NEC.

GROUNDING OF THE SERVICE EQUIPMENT

Proper grounding depends largely upon your bonding and coupling techniques. The service equipment is properly bonded

Table 1-3. OSHA Check List—Grounding

(Write Yes or No, Describe Equipment, Location)

CIRCUIT AND SYSTEM GROUNDING

- Is there a two-wire DC wiring system in plant?
- Is there a three-wire DC wiring system in plant?
- Are all the AC wiring systems in the plant secondary AC type systems?
- Does the maximum voltage to ground on the AC system exceed 150 volts?
- Are there service conductors supplying the AC system which are not insulated?
- Are there electric furnace circuits in the plant?
- Are there cranes operating over combustible fibers in Class III hazardous locations which are grounded?
- Are there circuits of less than 50 volts in plant?

LOCATION OF GROUND CONDUCTORS

- Are there locations where an objectionable flow of current occurs over the grounding conductor?

ENCLOSURE GROUNDING

- Are there service conductor enclosures such as service raceways, service cable sheaths, or armoring, when made of metal, which are not grounded?
- Are there other metal, ungrounded conductor enclosures?
- Are there metal enclosures within 6 feet of lightning rods?

EQUIPMENT GROUNDING

(NOTE: Exposed, non-current carrying metal parts of fixed equipment, which are liable to become energized,

(c) In a hazardous location?
(d) The motor operates with any terminal at more than 150 volts to ground?

- Are there controller cases for motors, except lined covers of snap switches, which are not grounded?
- Is there electric equipment on cranes which is not grounded?
- Are there electric signs and associated equipment which are not grounded?
- Are those signs accessible to unauthorized persons, uninsulated from ground and and from other conductive objects?
- Are there switchboard frames and structures supporting switching equipment which are not grounded?
- Are there direct-current single-polarity switchboards in the plant?

(NOTE: The metal parts of the following pieces of non-electric equipment must be grounded.)

- Are there frames and tracks of electrically-operated cranes which are not grounded?
- Is there equipment in the plant carrying in excess of 750 volts between connectors?

EQUIPMENT CONNECTED BY CORD AND PLUG

(NOTE: Exposed non-current carrying metal parts of cord and plug connected equipment, which are liable to become energized must be grounded.)

- Is there equipment in hazardous area not grounded?
- Is there equipment operated at more than 150 volts to ground which are not grounded, except:
  (a) Motors where guarded?

must be grounded.)

• Is there equipment supplied by means of metal clad wiring?

• Is there equipment located in wet locations and not insulated?

• Is there equipment located within reach of a person who can make contact with any grounded surface or object?

• Is there equipment located within reach of a person standing on the ground?

• Is there equipment in a hazardous location?

• Is there equipment in contact with metal or metal lath?

• Is there equipment operating with any terminal at more than 150 volts to ground?

(a) Are there enclosures for switches or circuit breakers in locations accessible to unqualified persons?

(b) Are there metal frames of electronically heated devices which are not permanently and effectively insulated from ground?

(c) Are there transformers mounted on wooden poles at a height of less than 8 feet from the ground?

(NOTE: Make sure the non-current carrying metal parts of the following types of equipment, regardless of voltage, are grounded. Describe location, how grounded.)

• Are the frames of stationary motor not grounded where the following conditions exist:

(a) Supplied by means of metal-enclosed wiring?

(b) Located in a wet place and not isolated or guarded?

(b) Those metal frames of electrically-heated appliances which are exempted?

• Is there equipment of the following types which are not grounded:

(a) Refrigerators?

(b) Air conditioners?

(c) Vending equipment?

(d) Portable, hand held, motor operated tools (drills, clippers, sanders, scrubbers, saws)?

(e) Cord and plug connected appliances used in wet or otherwise conductive locations, or by persons standing on the ground or on metal floors?

• Are there electrically-heated appliances, either portable, stationary or fixed, which have ungrounded metal frames?

SPACING FROM LIGHTNING RODS

• Is there equipment with metal frames and enclosures of electrical equipment within 6 feet of lightning rod conductors?

METHOD OF GROUNDING

• Is the method of grounding in violation of any of the following:

(a) Effective grounding? The path to ground from circuits, equipment, and conductor enclosures must be permanent and continuous. It must have ample capacity to safely conduct any currents liable to be imposed on it. It also must have an impedance sufficiently low to limit the potential above ground and to facilitate the operation of overcurrent devices in the circuit.

/more/

Table 1-3. OSHA Check List—Grounding (concluded)

(b) Common use of grounding conductor? The grounding conductor of a wiring system is used for grounding equipment, conduit, and other metal raceways or conductor enclosures—including service conduit or metal sheath and service equipment. (NOTE: The grounding connection may be made to a grounded cold water pipe near the equipment.)

(c) Equipment on structural metal? Electrical equipment secured to the grounded metal frame of a building is a valid ground.

METHOD OF BONDING

• Is your method of bonding in violation of any of the following:

(a) Threadless couplings made up tight for rigid metal conduit and electrical metallic tubing?

(b) Bonding jumpers meeting the other requirements of this section? Bonding jumpers are used around concentric or eccentric knockouts which are punched or otherwise formed so as to impair the electrical continuity to ground.

(c) Devices other than locknuts and bushings approved for this purpose?

BONDING AT THE SERVICE EQUIPMENT

• Is electrical continuity of the grounding circuit assured in these areas:

(a) Service raceways or service cable armor or sheath (except as provided in section 230-63(b) and section 250-55 of the NEC)?

motors, fixed or portable lamps, or other utilization equipment, lighting fixtures, cabinets, cases, or conduit that is not grounded as specified in 1910.314?

• Are there any locknut-bushing and double-locknut types of contacts that are depended upon for bonding procedures? (Bonding jumpers with proper fittings or other approved means must be used. This requirement applies to all intervening raceways, fittings, boxes, and enclosures between hazardous areas and the point of grounding for service equipment.)

• If flexible conduit is used where permitted, is there any conduit that is without bonding jumpers and proper fittings?

• Where there is an ungrounded service conductor of a wiring system in a Class I location that is supplied from an overhead line, is it not protected by an approved lightning protective device?

• Is this lightning protective device connected to service conductors in any way except on the supply side of the service disconnecting means, and is it bonded to the raceway system in any place except at the service entrance?

• Is there any general purpose type enclosure (where permitted) that is not provided with lightning protective devices and fuses?

LIVE PARTS

CLASS I DIVISION I AND II

• Are there any exposed live parts?

(b) Service equipment and enclosures containing service entrance conductors—including meter fittings and boxes—interposed in the service raceway or armor?

(c) Any conduit or armor which forms part of the grounding conductor to the service raceway?

• Is electrical continuity at the service equipment assured by:

(a) Bonding the equipment to the grounded service conductor?

(b) Making the couplings and threaded bosses on enclosures with joints wrench-tight?

HAZARDOUS AREAS
CLASS I DIVISION I AND II

• Is there any exposed non-current carrying metal part of equipment such as frames or metal exteriors of

• Concerning wiring in a Class I location, when supplied from a grounded AC supply system in which a grounded conductor is part of the service:

(a) Is the grounded service conductor not bonded to the raceway system or to the grounding conductor for the raceway system?

(b) Is the bonding connection to the grounded service conductor made on any other than the supply side of the service disconnecting means?

• Concerning wiring in a Class I location, when supplied from a grounded AC supply in which no grounded conductor is part of the service:

(a) Is there no metallic connection between the supply system ground and the raceway system at the service entrance?

(b) If there is a metallic connection, is its ampacity less than 1/5 that of the service conductors?

(c) Is it smaller than #10 (if of soft copper) or #12 (if of medium or hard-drawn copper)?

at the grounded service connector. The threaded couplings and threaded bosses must be made up wrench-tight wherever rigid conduit is involved. In addition, bonding jumpers should be used around concentric knockouts which are punched or otherwise formed to impair the electrical connection to ground. Never use locknuts or bushings for this purpose.

Enclosures for service equipment are grounded unless the voltage is less than 150 volts to ground, isolated from conducting surfaces, and not exposed to contact with a conducting surface.

Service raceways, cable armor, and the metal sheath of service cables must be grounded. Conduit and metal pipe from an underground supply must be grounded where lead-sheathed cable is bonded to the underground cable system.

Flexible metal conduit—when used between two sections of raceway, or between the end of the raceway and a service entrance enclosure—must always have a bond between the two "interrupted" sections. Make sure the jumper is of adequate size and it's attached by means of pressure connectors or other approved means. Install the jumper and attachments where they are protected against likely physical damage.

INSTRUMENT TRANSFORMER CIRCUIT GROUNDING

The secondary circuits of current and potential instrument transformers must be grounded where the primary windings are connected to circuits of 300 volts or more to ground. Switchboards must be grounded irrespective of voltage unless the primary windings are connected to circuits of 750 volts or less and have no live parts exposed to unauthorized personnel.

When a frame or case of an instrument transformer is accessible to other than qualified personnel it must be grounded. It need not be grounded if the primaries are not over 150 volts to ground and the unit supplies current only to meters.

If the windings and working parts operate at 300 volts or more to ground and are accessible to other than qualified personnel, grounding is necessary. But grounding is not necessary if the primaries of the current transformers are less than 150 volts to ground and are used exclusively to supply current to meters.

Cases of instruments, meters and relays operating at 750 volts or less must be grounded if:

• They operate with winding or working parts (not located in a switchboard) at 300 volts or more to ground, and are accessible to other than qualified personnel.

• Operated from current and potential transformers, or connected directly in the circuit on switchboards with or without live parts on the front of the panel.

Do not ground the cases if they are operated from current or potential transformers, or connected directly in the circuit on switchboards having exposed live parts on the front panels. In this case provide the operator with an insulated floor mat, if voltage to ground exceeds 150 volts.

The case of an instrument, meter or relay that operates at over 750 volts to ground must be isolated by elevation or protected by suitable barriers, grounded metal, or insulated covers and guards. The cases, themselves, need not be grounded. The internal ground segments of the instruments (except in electrostatic ground detectors) are connected to the instrument case and grounded; the ground detector is isolated by elevation.

GROUNDING METHODS THAT GET APPROVAL

• The path to ground from circuits, equipment, or conductor enclosures must be both permanent and continuous. It must have ample current carrying capacity to safely conduct any currents likely to be imposed on it. And importantly, it must have an "impedance sufficiently low" to limit the potential above ground and to facilitate the operation of the over-current devices in the circuit.

"Impedance sufficiently low" means that the grounding path and the circuit conductors are in the same metallic enclosure. This enclosure under certain conditions may be used to provide the equipment grounding path for the circuit conductors within it.

• The grounding conductor for a wiring system is also used for grounding equipment, conduit and other metal raceways or enclosures for conductors (including service conduit, cable sheath and service equipment). The grounding connection may be made to a grounded cold water pipe near the equipment.

• Electrical equipment secured to the ground structural metal frame of a building is considered to be grounded.

• Metal car frames supported by metal hoisting cables attached to or running over sheaves and drums of elevator machines are considered grounded when the machine itself is properly grounded.

• No objectionable passage of current can be present over the grounding conductors if the grounding of wiring systems, circuits, arrestors, cable armor, conduit or metals is to serve as a protective measure. (Temporary currents set up under accidental conditions while the grounding conductors are doing their intended protective function is not considered objectionable.)

• The use of multiple grounds may cause an objectionable flow of current over a grounding conductor. If that is the case, one or more of the grounds must be abandoned or their location changed, or the continuity of the conductor between the grounding connections must be interrupted or the current limited.

GROUNDED ITEMS SURE TO BE CHECKED

One of the inspector's favorite items to check for proper grounding is the office water cooler. This reason is simple. Many of the new model replacement coolers use plastic pipe instead of the "old-fashioned" copper. The copper offered a reliable, automatic, ground. Plastic pipe does not. It can become a shock hazard when placed next to a steel column. Run a separate ground wire to the cold water pipe and ground it properly. Problems occur in the metal-pipe water coolers because of the necessary calking. Check them, too.

Another frequently checked item is portable power tools. Make sure they all use three-wire type cords. If they are double-insulated (internally grounded), they will pass an OSHA inspection, but they must be clearly marked.

Lamps on machinery must be grounded. It is considered a valid ground if they are securely attached to a grounded machine. But many times the operator will place a piece of rubber padding, cloth, or other insulative material to tighten the lamp mount and stop vibration. This removes the ground.

All of your receptacles must, of course, be three-wire grounding type. Check them with a polarity meter. Simply insert it and read the light indications.

USE CORRECT BONDING AND GROUNDING FITTINGS

The NEC demands that certain basic practices be followed in making the electrical grounding system safe. Electrical fittings should meet these requirements:

- Must be UL-listed.
- Should be rugged, strong, and well-plated so that they will fasten and stay tight. They should not be sheet metal type ground straps.
- They must fasten mechanically. The NEC repeatedly emphasizes that solder should never be used for grounding connections.
- They must have the capacity for a large enough ground wire.
- They must be compatible with metals used in the system. For example, aluminum conductors should not be connected to copper connectors. Compatible aluminum conductors are available.

OSHA: – CHECKING OUT OVERCURRENT DEVICES AND TRANSFORMERS

Periodic inspections, during installation and continuously during service are the best way to assure OSHA compliance for overcurrent devices and transformers. These check-outs are not only a documented proof of "intent to comply," but they ensure the reliability of the equipment.

REVIEW THESE IMPORTANT CONSIDERATIONS

All electrical systems require overcurrent protection from fuses or circuit breakers. In 1974, OSHA required ground fault devices on New Construction sites.

Check interrupting capacity of all devices to be sure capacity is sufficient for the highest voltage and current to which it might be subjected. If, for example, the maximum short circuit current possible is 50,000 amperes, then the overcurrent device must be able to withstand this and clear the circuit without rupture.

Consider ambient temperature when installation is in a foundry, near an oven, or simply subjected to the direct rays of the sun on a hot day. Without proper ventilation, high temperatures distort the calibration. When normal in-plant ventilation proves inadequate, consult your equipment manufacturer for further suggestions.

Be sure overcurrent devices are free to open in case the circuit gets overloaded. This can be accomplished by using a trip-free or multiple breaker with one operating handle per pole. Even though normal operation is done by electrical or pneumatic power, an operator must be able to open and shut the device by hand for normal maintenance. There should be a clear indication whether the device is open or closed.

MOUNT DEVICES WITHIN EASY REACH

For the sake of emergencies as well as maintenance safety, devices must be mounted within easy reach. The common practice of mounting fusible switches up-high, out-of-the-way near overhead equipment is a clear violation of OSHA. Remount them so servicemen don't have to climb over obstacles or stand on a chair or portable ladder.

Although the NEC specifies no "maximum approved height," if the center of the enclosure is 6½ feet above the floor an OSHA inspector will usually approve it. This height also fulfills the requirement of keeping men safely away from suddenly moving parts such as the circuit breaker lever.

Keep any live parts of the unit away from combustible materials. Consider, in addition, that the unit must not be exposed to physical damage (such as too near aisleway, exposed to gases in the air, or subjected to a water drip).

THERE ARE SOME EXCEPTIONS

Locate overcurrent devices where the conductor to be protected receives its supply, except:

• When service overcurrent devices must be an integral part of the service disconnecting means, immediately adjacent, or at the outer edge of the entrance.

• When an overcurrent device protecting the larger conductors also protects smaller conductors in accord with Tables 310-12 to 310-15.

• Taps to individual outlets and circuit conductors supplying an electric range are considered protected if in accord with Sections 210-19 and 210-20.

• Conductors tapped from a feeder are considered properly protected when installed in accord with Sections 364-8 and 430-59.

• Conductors tapped to a feeder, where:

Length of conductor is less than 10 feet.

Ampacity of tap conductor is not less than the combined total loads of the circuits supplied by the tap conductors, and not less than the ampere rating of the switchboard or panelboard supplied by the tap conductors.

Protrusion of tap conductor is not beyond the switchboard.

• Feeder taps not over 25 feet long. Where the smaller conductor has ampacity at least one-third that of the conductor from which it is applied and provided the tap is protected from physical damage and terminates in a single circuit breaker or set of fuses which limit the load on tap. See Tables 310-12 to 310-15. Beyond this point, conductors may supply any number of circuit breakers or set of fuses.

• Transformer feeder taps with primary plus secondary not over 25 feet long. All the following must be met:

Conductors supplying the primary of a transformer have a capacity of at least one-third that of the conductors or overcurrent protection from which the primaries are tapped.

Ampacity of the conductors supplied by the secondary transformer, multiplied by the ratio of secondary-to-primary voltage, is at least one-third the ampacity of the conductors or overcurrent protection from which the primary conductors are tapped.

Length of the primary plus secondary conductor (excluding any portion of the primary conductor that is protected at its ampacity) is not over 25 feet.

Termination of secondary conductors in a single circuit breaker or set of fuses which limit the load allowed. See Tables 310-12 to 310-15.

PROVIDE ENCLOSURES FOR OVERCURRENT DEVICES

Enclose your overcurrent devices in a cutout box or cabinet, or in a specially-approved assembly offering equal protection. OSHA approves mounting them on switchboards, panelboards, or controllers as long as the room is free from ignitable materials and dampness. Maintain at least ¼-inch air space between enclosures and wall surface.

Damp or wet locations require a weather-tight enclosure.

Mount the box vertically, if at all possible.

Never use rosettes to mount fuses.

Whenever you enclose these devices, you must restrict their capacity to 80% of the rating for continuous load (except for devices installed on services operating at 600 volts). Where a vault of metal-enclosed switchgear is not used, fulfill OSHA requirements using:

• Air load interrupter switches (or other approved switches capable of interrupting the load) with suitable fuses on a pole or elevated structure outside the building. Device must be operable from within the building.

• Automatic trip circuit breaker (on circuits of any voltage) with an overcurrent device in each ungrounded conductor. Locate it on a pole, roof, foundation or other structure outside the building. It must be as near as possible to where service conductors enter.

The most frequently violated standard concerning overcurrent protection is locating the devices near combustible materials or exposing them to damage. But sometimes, as in the case of bus bars at the rear of a panel or switchboard, the only

Table 1-4. OSHA Check List—Overcurrent Devices

(Write Yes or No, Describe Equipment, Location)

PROTECTION OF PERSONNEL

• Are fuses or circuit breakers located where personnel can accidentally contact a live or suddenly moving part?

• Is the working space around the front of your equipment at least 30 inches? (NOTE: Working space is the distance between exposed live parts, or enclosure opening and nearest grounded surface.)

• Is any overcurrent device in the vicinity of an easily-ignitable material?

• Are fuse boxes on the control circuit placed where the maintenance electrician may be injured from them?

PROTECTION OF EQUIPMENT

• Are the circuit breakers (such as used for the branch circuits described in Section 210) not designed so any alteration to the trip point calibration or trip time will be difficult?

• In the case where condensation is a problem, are you using a space heater to clear it?

• Are there more than 42 overcurrent devices (other than those provided in the mains) in any one cabinet for a lighting and appliance branch circuit panel-board? (NOTE: A two-pole circuit breaker is considered as two devices, a three-pole as three devices, etc.)

• Does the load on any overcurrent device located in an enclosure exceed 80% of its rating, where in normal operation the load continues for at least three hours?

• Are all your overcurrent devices provided with appropriate enclosures for the hazards to which they might be subjected?

MARKING AND LABELING

• Is it clearly indicated on any overcurrent device whether open or closed?

• Are any ratings of a fuse or circuit breaker not clearly shown? (NOTE: It is permissible to have to remove a trim or cover to find the identification.)

• Is the ampere rating of circuit breakers rated 100 amps or less and 600 volts or less molded, stamped, or etched on the handle or "shield" of the device?

• Is the interrupting capacity of a breaker rated 240 volts or less and 100 amps or less (with an interrupting capacity other than 5000 amps) shown on the circuit breaker or on an attached label?

• Does each circuit breaker rated more than 240 volts or more than 100 amps (with an interrupting capacity other than 10,000 amps) show its rating on the circuit breaker or on an attached label? (NOTE: Interrupting ratings may be omitted on circuit breakers used only for supplementary overcurrent protection.)

way to offer protection is through the use of a permanent guard or fence. Locking up the enclosure discourages tampering by unauthorized personnel, and it also discourages "temporary overloaded circuit hookups" since the electrician knows the box will be permanently closed.

Normal maintenance procedures should never expose an employee to shock hazards. Not only must the maintenance technician be properly educated in the use of safety equipment (insulated blankets, gloves, hot sticks, etc.), but the area itself must be as safe as possible. Guards must be placed around fuse blocks. Space heaters must be used to rid condensation from the area. Two-bolt lugs must be used on cables 350 MCM and larger.

Although modern switchgear and overcurrent devices are properly designed, manufactured, and installed, malfunctions still occur due to inadequate preventive maintenance, lack of proper coordination between protective devices and loads, wrong calibration, and improper testing.

The plant's maintenance staff should test the equipment periodically for relay calibration, circuit breaker timing, and instrument calibration. It can be quickly determined if the devices are becoming frozen, pitted, or incorrectly aligned.

CHECK FUSES AND FUSEHOLDERS

When a fuse does blow, first check for evidence of tampering or overfusing. The rating of the device will usually be clearly stated on the case or on an attached label, but sometimes you may have to remove a trim or shield piece to read it.

Trip the circuit breaker and use a fuse puller (not your hands) to remove the burned-out device. Replace it with one of equal value and type, making sure you haven't left any live part of the fuse or holder exposed.

When renewing a fuse, keep in mind:

• Never use an Edison-base type plug fuse or a type S fuse in circuits rated over 125 volts, 30 amperes.

• Something's probably been tampered with if it is easy to change fuses of one type with another.

• Cartridge type fuses have no 250 volt ratings over 600 amperes, but a 600-volt fuse may be used for lower ratings.

• Make sure interrupting rating, ampere rating (where other than 10,000 amperes), voltage rating, manufacturer's name or trademark, and if "current-limiting" or not is clearly shown. Interrupting rating may be omitted on fuses used for supplementary circuit protection.

• Provide disconnecting means on the supply side of all your fuses (or thermal cutouts) in circuits of more than 150 volts to ground, and for cartridge fuses of any voltage where they are exposed to unauthorized persons. Except as described in Section 230-73, each fused circuit can be disconnected from the power supply.

• Always use a disconnecting means whenever cartridge-type fuses are required in panels. Check older installations for this, since it's a common violation.

• Supplementary overcurrent protection is often utilized in connection with appliances or utilization equipment to provide protection for specific components (or internal circuits within the equipment). This does not change the requirements of branch circuit protection, but these supplemental overcurrent devices do not have to be as accessible as branch overcurrent devices.

KEEP VOLTAGE TRANSFORMERS IN COMPLIANCE

Voltage transformer units that ought to be carefully inspected are found in the utility vault, load-centers throughout the plant, and at various subdistribution points where voltage is stepped down to utilization levels.

A quick look determines if the equipment is as free as possible from physical damage. This means keep it away from aisle-ways where a passing forklift might brush it. Repair the roof if rain is leaking in, and keep any gases or spray processes used in the plant away. Also, never pile materials on top.

Mountings should be absolutely secure so there is no chance of it falling over, or being moved. Other pieces of equipment must be kept away if they might interfere with it electrically.

Locate the unit so any maintenance or emergency repairs can be done in safety. This means relocate it if personnel have to climb a portable ladder or stand on a chair.

TRANSFORMERS INSTALLED INDOORS

Dry-type transformers, 600 volts or less, located in the open walls, columns, or structures, do not have to be accessible— nor do dry-type transformers below a rating of 600 volts and 50 kva when installed in a fire-retardant hollow space of a building (if well-ventilated and not permanently closed-in by the structure). Open-ventilated type transformers need the cleanest and driest air possible. Be especially careful not to allow water leaks in the transformer vault roof.

When these units have a rating of 112½ kva or less, they must have at least 12 inches of separation from combustible materials. Spacing may be less if a fire-retardant shield is used, or if the rating is less than 600 volts, and it's completely enclosed except for a vent opening.

Units rated more than 112½ kva must be installed in a fire-resistant transformer room unless constructed with 80C rise (Class B) or 150C rise (Class H) insulation and separated from combustible materials by not less than 6 feet vertically and 12 feet horizontally. Lessen this space if a fire-resistant, heat-insulating barrier is used.

Units rated more than 35,000 volts must always be installed in a vault. All outdoor installations must always use an approved, non-combustible, moisture-proof case that does not interfere with live parts.

An *askerel-insulated unit* rated in excess of 25 kva must be furnished with a pressure relief vent. If this unit must be installed in a poorly vented place, furnish it with a means of absorbing any gases generated by arcing inside the case. It's permitted by OSHA to connect the pressure relief vent to a chimney or flue that will carry gases outside the building. As with the dry-types, units rated over 35,000 volts must be put in a vault.

Always install *oil-insulated transformers* in a vault, unless:

• Not over 112½ kva capacity. Vault must be approved according to the NEC. The exception is when it is constructed of over 4-inch thick reinforced concrete.

• Not over 600 volts. A vault is not needed if suitable safeguards are provided against transformer oil fires from igniting other materials. In this case, total capacity of the transformer must not exceed 10 kva in any section of the building classified as combustible, or 75 kva where a surrounding structure is classified as "fire-resistant construction."

• An electric furnace transformer with a total rating not exceeding 75 kva may be installed without a vault if it is put in a building or room with fire-resistant construction. In this case, make sure a transformer fire cannot spread to combustible materials.

• Detached buildings. Transformers may be installed in a building which does not conform with NEC provisions for transformer vault providing neither the building nor the contents present a fire hazard to any other property. Make this building accessible only to authorized personnel, and use the building only for supplying electrical service (not storage).

Secondary grounding of indoor dry-type transformers was never before clearly spelled out. Now, OSHA insists you follow Section 250-26 of the 1971 NEC, retroactively.

TRANSFORMERS INSTALLED OUTDOORS

Whenever oil-insulated transformers are installed outdoors, make sure provisions are made to safeguard combustible materials, combustible parts of the building and its contents, fire escapes, and door and window openings from fires originating in transformer. According to the extent of the hazard, separate by space, put in fire-resistant barriers, install automatic spray systems, and/or enclosures which confine the oil to a ruptured transformer tank.

Oil enclosures should consist of fire-resistant dikes, curbed areas, basins or trenches filled with coarse stone. Install a trapped drain in cases where exposure and quantity of oil make removal necessary.

Tests on oil-filled equipment should include dielectric strength, acidity, interfacial tension, color and power factor.

If the oil is found to be unsuitable, it can either be reclaimed or filtered. In the reclamation process, moisture and

Table 1-5. OSHA Check List—Transformers

(Write Yes or No, Describe Equipment, Location)

GUARDING

• Are there transformers located where they are exposed to physical damage?

• Are there transformer installations which do not conform to the guarding provisions for live parts in Section 1910.310 (j)?

• Are there transformer installations which do not offer protection against accidental insertion of foreign objects?

• Are there transformers which do not have their exposed live parts indicated by markings on the equipment or structures?

• Is there any dry-type transformer located in a Division 2 hazardous area which does not have its windings and terminal connections enclosed in a tight metal housing without vents or other openings?

MARKING

• Are there transformers which are not provided with a nameplate giving:
  Manufacturer's name?
  Rated kilo-volt-amperes?
  Frequency?
  Primary and secondary voltages?
  Amount and kind of insulating liquid (where used) when rating exceeds 25 kva?
  Temperature rise of insulating system where Class B insulation is used for a dry-type transformer rated over 100 kva?

DRY-TYPE TRANSFORMERS INSTALLED INDOORS

• Are there any dry-type transformers installed indoors (rated 112½ kva or less, and more than 600 volts) which are not separated from combustible materials by

• Is there any transformer which may, when functioning at full rating, develop surface temperatures high enough to cause extensive dehydration of any organic dust deposits that may occur? (Maximum surface temperatures under operating conditions must not exceed 120C for equipment subject to overloading.)

• Are any transformer units installed where dust from magnesium, aluminum, aluminum-bronze, or other volatile metals are present?

GROUNDING

• Are there transformer installations which have exposed non-current carrying parts (including fences and guards), which are not grounded where required in the manner prescribed for exposed metal parts in Section 1910.314?

at least 12 inches?

• Are there dry-type transformers not provided with a non-combustible, moisture-resistant case or enclosure which offers reasonable protection from accidental insertion of foreign objects?

• Are there dry-type transformers of more than 112½ kva rating with more than 35,000 volts?

ASKAREL-INSULATED TRANSFORMERS

• Are there askarel-insulated transformers installed outdoors?

OIL-INSULATED TRANSFORMERS

• Are there oil-insulated transformers installed:
  Indoors?
  Outdoors?

acids are removed from the oil and the interfacial tension properties are restored. If testing shows the oil is simply wet or dirty, mechanical filtration is sufficient to restore it to usable condition.

PROVISIONS FOR TRANSFORMER VAULTS

Locate vault where it can get outside air without using flues or ducts if at all possible. The roof, walls, and floor can be constructed of reinforced concrete, brick, load-bearing tile, concrete block or similar fire-resistant material of adequate structural strength.

Vault floors in direct contact with the earth must be composed of at least 4-inch thick concrete. But when constructed with a vacant space or other stories under it, the floor is only required to have sufficient structural strength. A fire-resistance of at least 3 hours is required for floors, walls, and roof, according to Standard E119-67 (Fire Tests of Building Construction and Materials, NFPA No. 251-1969).

Vault doorways leading into the building must have tight-fitting doors of the Class A type (NFPA Standard for the Installation of Fire Doors and Windows). Some OSHA inspectors require this type door on an exterior wall opening on each side of an interior wall opening where conditions warrant. Call local OSHA headquarters for advice if it is thought that an added safety factor is required.

Sills or curbs must be high enough to contain within them all the oil from the largest transformer should it burst. In no case should this sill be less than 4-inches high. If a new transformer is installed with a larger oil capacity in the old vault, chances are good that it is out of compliance due to the sill height.

Locks on entrance doors should allow access only to qualified personnel. The lock must be one that can be opened quickly from the inside should the door close.

Ventilation openings should never be located near doors, windows, fire escapes, or combustible materials. If the equipment is vented by natural circulation, all openings should be placed in or near the roof, or half near the floor and the remainder in the roof or sidewalls near the roof.

If it is vented to the outside and flues or ducts are not used, the combined net area of all the openings must be at least 1 sq. ft. for any capacity under 50 kva.

Cover vent openings with durable gratings, screens, or louvers, according to the treatment required. Automatic closing dampers must shut the vent openings in response to a fire. They should not be less than No. 10MSG steel and have fire-retardent properties.

Drainage is recommended whenever possible. Vaults containing more than 100 kva capacity must be provided with a drain capable of carrying off oil or water accumulation. Pitch the floor toward the drain.

Water pipes and accessories foreign to the electrical installation should not be in the vault. Never store anything in the vault. (Facilities such as piping for fire protection or for the water-cooled transformers, or equipment to rid condensation, leaks, and breaks in the systems are not considered "foreign.")

MAKE SAFETY PROVISIONS FOR PERSONNEL

The greatest concern around a transformer is safety to personnel. It's best to isolate the unit and to make it inaccessible to unauthorized personnel. But when it is not in a separate room, elevate the unit at least 8 feet off the floor. Guard rails properly protect a unit rated not over 600 volts. For safety, eliminate wet conditions.

The nameplate on the transformer (in a readily seen spot) must state manufacturer's name; rated kilo-volt-amperes, frequency; primary and secondary voltage; and the amount and kind of insulating liquid used (if any) when the rating exceeds 25 kva. If Class B insulation is used in the construction of dry-type transformers rated at more than 100 kva, the nameplate must indicate the temperature rise for the insulation system.

Ground is required for exposed non-current carrying metal parts of transformer installations including fences, guard rails, etc. Install them according to Section 1910.314.

Because secondary grounding has never before been clearly spelled out in earlier codes, there have been a variety of methods used to ground indoor dry-type transformers. Since OSHA now requires adherence to 250-26, maintenance engineers ought to carefully re-check their plant for compliance.

BRING SWITCHES, BOXES AND OUTLETS INTO OSHA COMPLIANCE

OSHA often cites workmanship in the installation and maintenance procedures used for electrical receptacles, plugs and connectors, disconnect switches, and all related enclosures. Although an OSHA inspector will dislike seeing any sort of poor quality or sloppy installation, he is empowered to cite this sort of carelessness if it was done after March 15, 1972.

Even though installations may be electrically correct, if the end result is a confusing switching set-up with unmarked switches, a conductor unnecessarily exposed, peeling insulation that needs retaping, a device not tightly screwed into position, or other examples of neglect or oversight, OSHA will issue a citation. The federal government is saying that work must not simply be in compliance with the NEC, but also must be "electrically clean" with workmanship done with pride.

This is the reason why the NEC can not be used entirely for a design and installation standard, even though OSHA is based upon the NEC. Try to exceed the NEC, if possible. An OSHA inspector will examine the installation "as a whole" rather than simply checking that each part is in compliance.

CHECKING PLUGS AND CONNECTORS

Plugs and connectors must have all current carrying parts—except the prongs, blades, or pins—completely enclosed for safety. Open-face devices and certain fiber disc attachment plugs will not pass an inspection.

"Dead front" design with all wiring done from the back is definitely approved and will be approved in the future. These devices are sealed-off from foreign matter, offer complete separation of the conductors to reduce short circuit potential across the terminal screws, prevent loose strands, and protect personnel from accidentally coming into contact with live parts. When new plugs and receptacles are needed, it may pay to purchase this better design, possibly saving a replacement cost in later years.

USE NEW CONFIGURATIONS

NEMA and the wiring device manufacturers have developed plug and receptacle configurations that differ with all ratings. This completely eliminates the possibility of hooking together the wrong units. These grounding type devices are offered in various twist-lock and straight-blade forms. OSHA has issued many citations because of improper use of the right configuration where locking devices have been installed, or where the use of crowfoot devices is prevalent. If either is the case in your plant, check the devices carefully.

Environmental conditions are also important. Even if correct devices are used, but they are sitting in a pool of water, a citation will surely be given. Also, if the device is for use at over 300 volts, it must have "skirts," to confine any possible arcing.

Many of the old-style configurations have been de-listed by UL. But if they are properly grounded and they comply with OSHA's intent of personnel safety, they may still be used on work that was in-place before March 15, 1972. They may also be used as replacements for this work.

But, consider that OSHA will get stronger as it continues developing and might eventually force you to replace these old-style devices.

MAKE SURE DEVICES COMPLETE GROUNDING CIRCUIT

OSHA requires all receptacles to have an operative grounding circuit. A self-grounding type receptacle insures this compliance, because this type of device is automatically grounded if installed in a grounded metal box. A heavy phosphor bronze

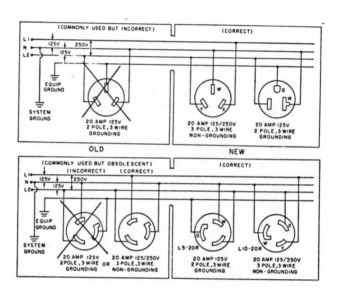

Figure 1-2. Receptacles

Shown above are incorrect and obsolescent designs and the correct configuration. On the facing page are shown new plug and receptacle configurations that differ with all ratings, eliminating the possibility of hooking
together the wrong units.

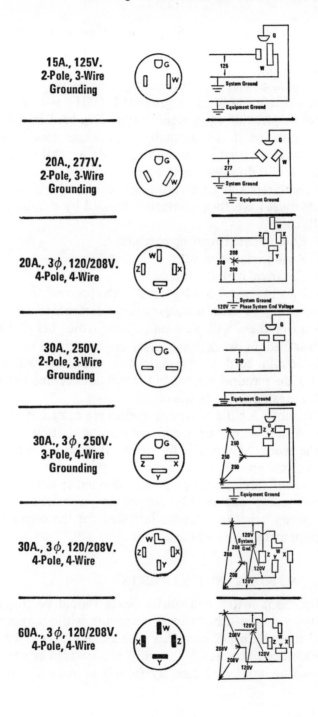

15A., 125V.
2-Pole, 3-Wire
Grounding

20A., 277V.
2-Pole, 3-Wire
Grounding

20A., 3φ, 120/208V.
4-Pole, 4-Wire

30A., 250V.
2-Pole, 3-Wire
Grounding

30A., 3φ, 250V.
3-Pole, 4-Wire
Grounding

30A., 3φ, 120/208V.
4-Pole, 4-Wire

60A., 3φ, 120/208V.
4-Pole, 4-Wire

grounding strap connects the ground. Another benefit is that no green jumper has to be installed (or has the possibility of coming loose).

All cord-connected appliances getting power from such an approved receptacle must be grounded. OSHA is sure to check any tool or piece of equipment that's being used in a damp or wet location, or if it's normally used while standing on the ground, or if it's used while working inside a metal tank or boiler. If the tool is double-insulated, it is probably in compliance.

Receptacles and cord connectors equipped with grounding contacts must have their contacts grounded by means of one of the following:

- Equipment grounding conductor.
- Rigid metal conduit.
- Electrical metallic tubing.
- Flexible metal conduit used with approved fittings.

Never try to make a nongrounding device into a grounding one. Replace them with new units. Also remember that the devices installed in work areas must endure rough use. A nonmetallic strain cord is recommended since this will prevent strain on the terminal screws, but it will not become conductive should a break develop in the cord.

Terminals for the grounded devices are color coded:

- White (identified by a white color coating or marked with the word "white") is the terminal intended for the connection of the grounded conductor.
- Green (identified by a green color and an octagon shaped terminal screw, not readily removable, or marked with the word "green") is the terminal intended for the connection of the equipment grounding conductor.

INSPECTING BOXES AND ENCLOSURES

Junction, switch and outlet boxes should be installed so that the wiring within them is accessible without removing a part of the building or paving. Fasten them rigidly and securely to the surface upon which they will be mounted or embedded. They can be supported directly from a structural part of the building or from an approved brace. OSHA calls for a wooden brace to be at least 1 inch thick, and a metal brace to be at least No. 24 MSG or 0.0239 inch thick.

Approved anchors or clamps may be used for new walls that have no structural members or in existing walls of previously occupied buildings. Use 100 cubic-inch size boxes and securely affix them. OSHA considers it adequate support for threaded boxes less than 100 cubic-inch size which do not contain a device if they are secured by two or more conduit on two or more sides of the box. These conduits must be threaded into the box wrenchtight and supported no more than 3 feet from the box.

Make sure outlet boxes for concealed work have an internal depth of at least 1½ inches. But if this larger depth will result in an injury or is impractical, use a shallower enclosure—providing it is at least ½ inch deep.

Check all connections inside and outside the box. If there is an exposed extension from an existing outlet to exposed wiring, make sure the box, extension ring, and blank cover are electrically and mechanically secure.

INSTALL PROPER BOX FOR SITUATION

• Floor-mounted devices must use a box designed for that purpose. Cast metal boxes may be used on sub-floor and ground-floor to meet the water-tight requirements. Stamped steel boxes may be used anywhere above ground level.

• Wall or ceiling mounted devices to be installed in concrete, tile, or other noncombustible material must be placed so the front edge of the box or fitting will not set back of the finished surface deeper than ¼ inch. When installing the devices in combustible materials, mount the box so it is either flush or protruding from the finished surface.

• Flush-mounted units must be installed in a box that completely encloses them on the back and sides, and offers substantial support for the device. The same screws that attach the device in the box cannot also support the box.

In general, boxes will be metal, but approved non-metallic boxes are sometimes used with open wiring on insulators, concealed knob-and-tube-work, non-metallic sheathed cable, and

Table 1-6. OSHA Check List—Outlet, Switch and Junction Boxes, and Fittings

ROUND BOXES
• Are there round boxes in use?

NON-METALLIC BOXES
• Are there non-metallic boxes in use? (NOTE: Double check the equipment installed in corrosive areas.)

METALLIC BOXES
• Are there metallic boxes used with knob-and-tube work or non-metallic sheathed cable, and mounted on metal lath ceilings or walls which are not insulated from their supports and not grounded?

DAMP AND WET LOCATIONS
• Are there boxes and fittings in damp or wet locations which are not weatherproof?

NUMBER OF CONDUCTORS IN BOX
• Are there boxes which are not of sufficient size to provide free space for all conductors enclosed in the box? (NOTE: These provisions apply to terminal housings supplied with motors. They are to be used as inspection criterion, but do not apply to conductors used for rewiring existing raceways.)
• Is the maximum number of conductors, including grounding conductors but not counting fixture wires, in outlet and junction boxes equivalent to Tables 370-6(a)(1 and 2) and 370-6(b).

(c) Are there raceways and cables installed with metal outlet boxes or fittings which are not secured to such boxes and fittings?

NON-METALLIC BOXES
• Are there non-metallic boxes used with open wiring or concealed knob-and-tube work, into which conductors do not enter through individual holes?
• Is there flexible tubing used to encase conductors which does not extend from the last insulating support either into a box or terminate at the wall of the box?
• Is there non-metallic sheathed cable (assembly) used which does not enter the box through a knockout?
• Are there conductors or cables entering a box which are not clamped to the box and which are not supported within 8 inches of the box?
• Is there non-metallic conduit installed with non-metallic boxes or fittings which is not secured to the boxes or fittings in a manner consistant with the preceding requirements?

370-6(a)(1) DEEP BOXES

Box Dimensions	Cubic Inch Max.		Number Conductors			
Inches, Trade Size	Capacity	No.	14	12	10	8
3¼x1½ Oct.	10.9		5	4	4	3
3½x1½ Oct.	11.9		5	5	4	3
4x1½ Oct.	17.1		8	7	6	5
4x2⅛ Oct.	23.6		11	10	9	7
4x1½ Sq.	22.6		11	10	9	7
4x2⅛ Sq.	31.9		15	14	12	10
4 11/16x1½ Sq.	32.2		16	14	12	10

Box Dimensions	Volume				
4 11/16x2⅛ Sq.	46.4	23	20	18	15
3x2x1½ Dev.	7.9	3	3	3	2
3x2x2 Dev.	10.7	5	4	4	3
3x2x2¼ Dev.	11.3	5	5	4	3
3x2x2½ Dev.	13.0	6	5	5	4
3x2x2¾ Dev.	14.6	7	6	5	4
3x2x3½ Dev.	18.3	9	8	7	6
4x2⅛x1½ Dev.	11.1	5	4	4	3
4x2⅛x1⅞ Dev.	13.9	6	6	5	4
4x2⅛x2⅛ Dev.	15.6	7	6	6	5

(NOTE: These tables apply where no fittings or devices such as fixture studs, cable clamps, hickeys, switches, or receptacles are contained within the box. Where one or more fixture studs, cable clamps, or hickeys, are contained in the box, the number of conductors shall be one less. Deduct one more conductor if device is flush mounted.

A conductor running through the box is counted as one conductor and each conductor originating outside the box and terminating inside is counted as one conductor. Conductors which do not leave the box are not counted. If single, flush mounted boxes are ganged and each section is occupied by a flush device or combination of flush devices on the same strap, the limitations will apply to each section individually.)

BOXES

• Are there boxes other than those described in the tables not durably marked with the manufacturer's name or trademark and cubic inch capacity?

CONDUCTORS ENTERING BOXES OR FITTINGS

• Are there conductors entering boxes or fittings which are not protected from abrasion?

(a) Are there opening through which conductors enter which are not adequately closed?

(b) Are there metal outlet boxes or fittings installed with open wiring or concealed knob-and-tube work into which conductors do not enter through insulated bushings or through flexible tubing?

370-6(a)(2) SHALLOW BOXES*

Box Dimensions	Max. Number Conductors		
Inches, Trade Size	No. 14	12	10
3¼	4	4	3
4	6	6	4
1¼x4 sq.	9	7	6
4 11/16	8	6	6

(*NOTE: Any box less than 1½ inches deep is considered to be a shallow box.)

370-6(b) VOLUME REQUIRED FOR PER CONDUCTOR

Size Of Conductor	Cu. In. Free Space Within Box For Each Conductor
No.	
14	2.0
12	2.25
10	2.5
8	3
6	5

/more/

Table 1-6. OSHA Check List—Outlet, Switch and Junction Boxes, and Fittings (concluded)

UNUSED BOXES

- Are there unused boxes and fittings which are not closed to afford protection substantially equivalent to that of the wall of the fittings or box?
- Are there metal plugs or plates used with non-boxes or fittings which are not recessed at least ¼ inch from the outer surface?

BOXES ENCLOSING FLUSH DEVICES

- Are there boxes used to enclose flush devices which do not completely enclose the back side of the devices and provide substantial support for them?
- Are there boxes in which the screws for supporting the box are also used to attach the device to the box?

IN WALLS AND CEILINGS

- Are there boxes and fittings so installed in walls or ceilings or concrete, tile, or other noncombustible material that the front edge of the box or fitting is set back from the finished surface more than ¼ inch.
- Are there boxes and fittings so installed in walls or ceiling of wood or other combustible material that the front edge of the box or fitting is not flush with the finished surface, or projects from the surface?

REPAIRING PLASTER

- Are there plaster surfaces except on walls or ceilings composed of concrete, tile, or other noncombustible material, which are broken or incomplete due to the installation of a box?

- Are there any wooden braces less than 1 inch thick?
- Are there boxes less than 100 cubic inches in size, that are mounted in walls, in which no structural members are provided?
- Are the members of the approved type and the installation rigid and secure?
- Are there threaded boxes or fittings less than 100 cubic inches in size (not containing devices or support fixtures or supporting conduit) that are not supported within 3 feet of the box on at least two sides?

DEPTH OF OUTLET BOXES FOR CONCEALED WIRES

- Are outlet boxes used for concealed work which do not have an internal depth of a least 1½ inches?

COVERS AND CANOPIES

- Are there outlet boxes which are not provided with a cover or a fixture canopy?
  (a) Are there metallic covers or plates used on non-metallic boxes which do not comply with the grounding requirements of 1910.314(d)(1)?
  (b) Are there locations where a fixture canopy or pan has been used and a combustible wall or ceiling finish is exposed between the edge of the canopy or pan and the outlet box?

EXPOSED EXTENSIONS

• Are there any exposed extensions from an existing outlet of concealed wiring which have not been electrically and mechanically secured to a box, extension ring, or blank cover mounted on the original box?

• Are there any extensions which are not connected to the extender box in the manner prescribed for the particular method of wiring employed in making the extension?

SUPPORTS

• Are there boxes which are not securely and rigidly fastened to the surface upon which they are mounted?

• Are there any metal braces not corrosion-resistant or not less than 0.0239 inch thick (No. 24 MSG)?

(c) Are there covers or outlet boxes having holes through which flexible cord pendants pass that do not have smooth, well-rounded surfaces on which cord might bear?

(NOTE: Where metallic covers are used, they shall comply with the grounding requirements in 1910.314(d)(1).)

ACCESSIBILITY OF JUNCTION, PULL, AND OUTLET BOXES

• Are there junction, pull, and outlet boxes so installed that the wiring contained in them may not be rendered accessible without removing any part of the building, sidewalk or paving?

with approved non-metallic conduit. Non-metallic boxes are sometimes equipped with a metal grounding plate which has screws or terminals for connecting the grounding conductors. But, if such a plate is not offered on the particular device, it's acceptable to splice the grounding conductors together (with an approved splice inside the box) and connect them to a metal part of the installation.

Make sure the grounding conductors inside a wiring box have good electrical conductivity with each other, with the grounding terminal of the device, and with the box itself (if metallic). If a switch or circuit breaker is used in a circuit of over 150 volts to ground and is accessible to other than qualified personnel, be sure the device is properly grounded.

Also, if metal boxes are used for knob-and-tube-work on non-metallic sheathed cable mounted on metal or metal lath ceilings or walls, ground the metal or metal lath and the box—or insulate the box completely from the grounded supports and from the grounded metal laths.

LEAVE ENOUGH SPACE FOR WIRING

All boxes should be clearly identified with the manufacturer's trademark and cubic inch capacity. The maximum number of conductors (including ground conductors, but not counting fixture wires) are shown in the OSHA Check List—Outlet, Switch and Junction Boxes, and Fittings. When these enclosures are used for switches and overcurrent devices, they can be used as auxiliary gutters or junction boxes providing the conductors do not fill up more than 40% of the cross sectional area of the space and no more than 75% of the cross sectional area of the space is filled up with conductors combined with splices and taps.

Make sure the conduit entering into a box is protected from abrasion. Where metal outlet boxes or fittings are installed with open wire or concealed knob-and-tube-work, the conductors must enter through insulated bushings. In dry locations, they can enter through flexible tubing extending from the last insulating support and firmly secured to the box or fitting.

Where raceway cable is installed with metal outlet boxes or fittings, secure the raceway or cable to the box.

Where non-metallic boxes are used with open wiring or concealed knob-and-tube-work, the tubing extends from the last insulating support and it runs into the box, or terminates at the wall of the box. Non-metallic sheathed cable enters a box through a KO opening. The individual conductors or cables need not be clamped to the box if they are supported less than 8 inches away. It's not advisable to use a round box in a location where conduit or connectors require the use of locknuts or bushings to be connected to the side of the box. These can never be tightened completely flush.

USE AN APPROVED, SNUG-FITTING COVER

Be sure boxes are fitted with snug-fitting, UL-listed covers. OSHA has already issued many citations for coverless, broken, or hanging covers. Ferrous metal covers must be at least 0.030 inch thick. Nonferrous metal covers must be at least 0.040 inch thick. Covers made of an insulating material such as plastic must be noncombustible and at least 0.10 inch thick. OSHA will allow a thinner version of plastic cover if it is reinforced for additional strength.

Covers for non-metallic outlet boxes can be of either non- or metallic composition. If metal is used, connect it tightly to insure a good ground.

OSHA has made particular reference to covers for boxes that get a certain type use. For example a flush-mounted snap switch mounted in an ungrounded metal box, within reach of a conducting floor or other surfaces, must use a nonconducting, noncombustible cover. The aluminum outdoor weatherproof receptacle with a self-closing cover is completely acceptable if it is used only to energize portable tools that would not be connected indefinitely. Any cover used in a damp or wet location, such as this, must not allow moisture to accumulate in/on box.

Placing the cover on a box does not mean you're finished. Close up any openings in the box with KO plugs designed for that use. This plug must afford the same protection as the original wall of the box and it should be installed with a recess at least ¼ inch from the outer surface.

Often a combustible wall or ceiling finish is exposed between the edge of the canopy and the outlet box after the installation. Cover with noncombustible materials for fire protection. If the installation is in plaster, fill up any gaps or open spaces that may have developed around the box or fitting. A repaired wall or ceiling must have the same degree of fire protection as the original.

INSTALL SWITCHES PROPERLY

Install the switch into the circuit so that the grounded conductor cannot be disconnected before the ungrounded conductor or conductors have been disconnected. It is in compliance if all conductors are disconnected simultaneously. It's important to recheck three- and four-way switches to be sure the switching is done only in these ungrounded circuit conductors. In cases where a metal enclosure is used, the wiring between switches and outlets must be run with polarities in the same direction.

All switches should be installed for maximum accessibility. This means the operating mechanism must be lower than 6½ feet off the floor and it should not be necessary to climb over obstacles or use a portable ladder. It is permissible to group them together in outlet boxes if the voltage between adjacent units is less than 300, or permanently installed barriers are used between the devices.

Inspectors will often accept *overhead busway tap switches* or *hook stick* or *pull chain* operated devices as being accessible. This depends upon the particular plant situation and the degree of safety these kinds of disconnects offer. Check with local OSHA Area Director before ripping out old work.

On some machine tool applications, the *disconnect switch* is deliberately set into a position not readily accessible. This may be to protect the repairman while he is working on the machinery. Or it may be placed underneath the machine so that he can conveniently start and stop the unit. Although the reasoning behind these particular situations is to insure worker safety, OSHA still wants a disconnect switch within sight of the operator, in his normal operating position.

This disconnect can be either completely enclosed or remotely operated. In addition to preventing shock upon contact, the device must not create a hazard if it arcs under full load. Locate it so the machine operator does not have to reach over a moving piece of machinery. Sometimes magnetic switches may be preferred. They have a high degree of safety since they have to be manually reset after they are disconnected.

Any switch installed in a hazardous area must meet the special requirements of that area. Check with the switch manufacturer and inquire what particular models they recommend for Class I, II, and III locations.

General-use snap switches are for general distribution and branch circuit use. They are rated in amperes and capable of interrupting their rated current at their rated voltage.

• Surface type snap switches used on open wiring on insulators must be mounted on sub-bases of insulating material. This separates the conductors at least ½ inch from the surface wired over.

• Box-mounted, flush-type snap switches (as permitted in 370-10) must be neatly installed so the extension plastic ears are seated against the surface of the wall. Mounting in this fashion prevents the device from moving.

General-use AC devices control:

• Resistive and inductive loads (including electric discharge lamps) not exceeding the ampere rating of the switch at the voltage applied.

• Tungsten-filament lamp loads not exceeding the ampere rating of the switch at 120 volts.

• Motor loads not exceeding 80% of the ampere rating of the switch at its rated voltage.

General-use AC/DC devices control:

• Resistive loads not exceeding 50% of the ampere rating of the switch at the applied voltage. Horsepower rated switches are suitable for controlling motor loads within their ratings at the voltage applied.

Table 1-7. OSHA Check List—Cord-Connected Appliances

(Answer Yes or No, Describe Equipment, Location)

LIVE PARTS

• Are there appliances in which live parts are normally exposed to contact?

MOTOR-OPERATED APPLIANCES

• Are there motor-operated appliances in use which do not meet the requirements of the NEC?

FLEXIBLE CORDS

• Are there appliances, connected with a flexible cord, which are not of the approved type?

• Are there electric heaters of the portable immersion type which are not constructed and installed so that current-carrying parts are effectively insulated from electrical contact with the substance in which immersed?

PROTECTION OF COMBUSTIBLE MATERIAL

• Are there electrically heated appliances, obviously intended by size, weight, and service to be located in a fixed position, which do not have ample protection from adjacent combustible material?

STANDS FOR PORTABLE APPLIANCES

• Are there any portable electrically heated appliances, intended to be applied to combustible material, which are not equipped with an approved stand?

SIGNALS FOR HEATED APPLIANCES

• Are there electrically heated appliances or groups of them which are intended to be applied to combustible material that do not contain an integral temperature device?

(NOTE 1: For fixed appliances rated at not more than 300 volts or one-eighth horsepower, the branch circuit overcurrent device can serve as a disconnect. For fixed appliances of greater rating the branch circuit switch or circuit breaker may, where readily accessible to the use of the appliance, serve as the disconnecting means.)

(NOTE 2: For portable appliances, a separable connector or an attachment plug and receptacle may serve as the disconnecting means. The rating of the receptacle or separable connector shall not be less than the rating of the connected appliance.

Plugs and connectors shall be constructed and installed to guard against contact with live parts. They must be capable of interrupting their rated output without hazard to operator, and they must be designed so they do not fit into receptacles of a lesser rating.

For cord-connected appliances such as freestanding household-type ranges, a separable connector or an attachment plug and receptacle may serve as the disconnecting means.)

(NOTE 3: Unit switches which are part of the appliance shall be considered as taking the place of the disconnecting means, unless there are other disconnecting means. The branch circuit switch or circuit breaker, where readily accessible to the user of the appliance, may be used for this purpose.)

(NOTE 4: Switches and circuit breakers used as a disconnecting means must be of the indicating type.)

WATER HEATERS

• Is there a water heater which is not equipped with temperature limiting means in addition to the control thermostat?

INFRARED LAMP INDUSTRIAL HEATING APPLIANCE

• Are there infrared heating lamps rated at 300 watts or less which are being used with lampholders not designed for that purpose?

• Are there appliances which are not provided with infrared lamps over 300 watt rating which have not been especially approved for that purpose?

• Are there lampholders being operated in series on circuits of more than 150 volts to ground which do not have a voltage rating equivalent to the circuit voltage?

(NOTE: Each section panel or strip carrying a number of infrared lampholders, including the internal wiring of each section panel or strip, is considered an appliance. The terminal connection block of each such assembly is deemed an individual outlet.)

GROUNDING

• Are there metal frames of portable, stationary, and fixed electrically heated appliances operating on circuits above 150 volts to ground which are not grounded in the manner specified by 1910.314?

DISCONNECTING MEANS

• Are there electrical appliances supplied by more than one electrical source?

• Are there appliances which are not provided with a means for disconnection from all underground conductors as required?

(NOTE 5: When a switch or circuit breaker is used as a disconnecting means for a stationary or fixed motor-driven appliance of more than one-eighth horsepower, it must be located within sight of the motor controller and be capable of being locked in the open position.)

HAZARDOUS AREAS

CLASS I DIVISION I AND II

• Are there any switches and circuit breakers, and make and break contacts of pushbuttons, relays, and alarm bells or horns that do not have enclosures approved for the location?

(NOTE: To facilitate replacements, process control instruments may be connected through flexible cord provided the answers to the following five questions are answered negatively.)

• Is current interrupted by pulling out the plug?

• Is the power supply cord longer than 3 feet?

• Is the power supply cord not approved for extra-hard usage?

(NOTE: Cords approved for hard usage are equivalent to cords approved for extra-hard usage provided that they are protected by location and are supplied through a plug and receptacle of the locking and grounding type.)

• Are there any unnecessary receptacles?

• Are there any receptacles which do not carry a warning about unplugging while under load?

• In locations where conduit enters an enclosure for switches, circuit breakers, fuses, relays, resistors or other apparatus which may produce arcs, sparks, or

/more/

Table 1-7. OSHA Check List—Cord-Connected Appliances (concluded)

high temperatures, are there any seals which are more than 18 inches from the enclosure?

• Are splices and taps made in any fittings which was intended only for sealing with compound?

• Are there other fittings (in which splices or taps are made) filled with compound?

• If there is any possibility that liquid or other condensed vapor may be trapped within the enclosure for control equipment or at any point in the raceway system, are means not provided to prevent accumulation or to permit periodic draining of such liquid or condensed vapor?

• Are there any switches, circuit breakers, motor controllers, or fuses (including pushbuttons, relays, and similar devices) that are not provided with enclosures? (The enclosure and enclosed apparatus must be approved for Class I locations.)

• Is there any receptacle or attachment plug that is not of the type providing for connection of the grounding conductor of the flexible cord?

• Is there any receptacle or attachment plug that is not approved for the location?

• Are there any exposed live parts on receptacles, attachment plugs, switches or boxes?

• Are there any fittings or boxes that are not provided with threaded bosses for connection to conduit or cable terminations?

• Are there any fittings or boxes that do not have close fitting covers?

• Is there any union, coupling, box or fitting in the conduit between the sealing fitting, and the point at which the conduit leaves the Division I hazardous area?

• Are there any enclosures for connections or for equipment that are not provided with approved integral means for sealing?

• Are there any sealing fittings being used that are not approved for Class I locations?

• Are any sealing fittings not accessible?

• Is the sealing compound approved for the purpose?

• Is the sealing compound affected by the surrounding atmosphere or liquids?

• Does the sealing compound have a melting point of less than 200F (93C)?

• In any completed seal, is the minimum thickness of the sealing compound less than the trade size of the conduit? (In no case shall the minimum sealing compound thickness be less than $5/8$ inch.)

• Are there any fittings or boxes that have openings (such as holes for attachment screws) through which dust might enter or burning material might escape?

• Are there any fittings or boxes in which taps, joints, or terminal connections are made (or which are used in locations where dusts are an electrically conducting nature), that are not dust-ignition-proof and approved for Class II locations?

• Where flexible connections are necessary, is there any location without dust-tight flexible connectors, flexible metal conduit with approved fittings, or flexible cord-approved for extra-hard usage and provided with bushed fittings?

• If dusts are electrically conducting, is flexible metal conduit used, or are flexible cords not provided with dust-tight seals at both ends?

• Are there any disconnecting or isolating switches (containing no fuses and not intended to interrupt current) that are not provided with effective means to prevent the escape of sparks or burning material?

• In locations where dust from magnesium, aluminum, aluminum-bronze powders or other metals of extremely hazardous characteristics may be present, are there any switches, fuses, motor controllers, or circuit breakers which do not have enclosures specifically approved for such locations?

FITTINGS AND SEALS

• Are there any junction boxes or similar enclosures in any conduit run between the sealing fitting and the apparatus enclosure?

• Is there any conduit run (2 inch size or larger) entering the enclosure, fitting housing terminals, splices or taps which has seals more than 18 inches from the enclosure?

• Tungsten-filament lamp loads not exceeding the ampere rating of the switch at the applied voltage, when "T" rated. (See Sections 600-2 for sign and outline lighting, and 430-83, -109, and -110 for controlling motors.)

Position *single-throw knife switches* so that gravity will not tend to close them. Connect the units so the blades are dead when in an open position. If a double-throw unit is used, it may be installed in a horizontal or vertical position—providing a locking device is used that keeps the blades open when set.

Renewable auxiliary contacts or the quick-break type or equivalent must be provided on all 600-volt knife switches to break current over 200 amperes. It is recommended that these contacts be provided on all switches rated over 250 volts.

Use knife switches in the following manner:

• *For Isolating Use.* They should be rated more than 1200 amperes at 250 volts or less, or for more than 600 amperes at 251 to 600 volts. They must not open under load.

• *Using Special Designs.* To interrupt currents greater than 1200 amperes at 250 volts or less, or 600 amperes at between 251 to 600 volts, use a specially designed switch or circuit breaker for that purpose.

• *For General Use.* Knife switches of a lower rating may be used as general-use switches and may be opened under load.

• *For Motor Control Use.* Motor control switches must be of the knife switch type. Rated in horsepower, these devices must be capable of interrupting the maximum operating overload current of a motor with the same horsepower rating as the switch at the rated voltage.

LOCKING AND TAGGING SWITCHES

If there is any doubt at all what purpose the switch serves, mark it clearly. For example, a switch on a piece of BX cable going from an outlet to a piece of machinery is quite clearly a disconnect for the machinery. Yet, if the cable goes underneath the floor, the switch would have to be labelled as a disconnect for the machine since it is not self-evident.

Whenever possible to do so, all circuits—even low voltage ones—should be de-energized before work is done on them. Just opening the switch is not enough. It should be locked open for safety. Then a tag should be placed on it to warn against closing the circuit. The person doing the work or his supervisor, should keep the key until the work is completed.

OSHA: – INSPECT, CHECK AND CORRECT WIRING VIOLATIONS

Checking the wiring circuits and related wiring installations is of the utmost importance to avoid citations from the OSHA inspector.

Keep in mind certain facts about the plant:
• Does the age of the building justify an especially close examination of the plant?
• Are there any new NEC regulations that were not yet the law when the plant was built?
• Has any equipment been added, removed, or moved without a re-examination of the wiring?
• Has there been a change in process methods, or the introduction of a new process that is hazardous?

START BY REFERRING TO THE NEC

Generally, if a plant is fairly new and built in compliance with the 1971 National Electrical Code, there is an excellent chance of passing an OSHA wiring inspection with only a minimum of problems.

Older plants may find some of the retroactive sections of the NEC troublesome, but not impossible. They apply to instructions on connections, splices, protection from live parts, grounding methods, use of flexible cords, and special requirements for hazardous locations. These sections are 110-114, -117, -118, -121, -122; 240-16, -19; 250-3, -5, -7, -42 through -45; 250-50, -57 through -59; 400-3 through -5, -9 and -10; 430-142 and -143; 500 through 503.

Table 1-8. OSHA Check List—Wiring Systems

(Answer Yes or No, Describe Equipment, Location.)

WIRING METHODS

• Is there any location where threaded rigid metal conduit or Type MI, MC or ALS cable (with approved termination fittings) is not employed?

• If Type MI cable is used, is it installed so that there is tensile strength at the termination fittings?

• Are there fittings or boxes that are not provided with threaded bosses for connection to conduit or cable terminations?

• Are there any fittings or boxes that do not have tight-fitting covers?

• Are there any fittings or boxes in which taps, joints, or terminal connections are made (or which are used in locations where dusts are of an electrically-conducting nature), that are not dust/ignition-proof and approved for Class II locations?

• Are there any fittings or boxes that have openings (such as holes for attachment screws) through which dust might enter or through which sparks or burning metal might escape?

• Where flexible connections are necessary, are there any locations without dust-tight flexible connectors, flexible metal conduit with approved fittings, or flexible cord approved for extra-hard usage and provided with dust-tight seals at both ends?

• If flexible cord is used, are you using an additional conductor for grounding?

• If flexible connections are subject to oil or other

supply side of the service disconnecting means and is it bonded to the raceway system at any place other than the service entrance?

• Concerning wiring in a Class I location, when supplied from a grounded AC supply system in which a grounded conductor is part of the service:

   (a) Is the grounded service conductor not bonded to the raceway system and also not to the grounding conductor for the raceway system?

   (b) Is the bonding connection to the grounded service conductor made on any other than the supply side of the service disconnecting means?

• Concerning wiring in a Class I location, when supplied from a grounded AC supply in which no grounded conductor is part of the service:

   (a) Is there a metallic connection between the supply system ground and the raceway system at the service entrance?

   (b) If there's a metallic connection, is the ampacity less than 1/5 that of the service conductors?

   (c) Is it smaller than #10 (if made of soft copper) or #12 (if made of medium or hard-drawn copper)?

• Are any wiring systems in Class II locations, when supplied from overhead lines, unprotected from high voltage surges (due especially from lightning)?

• Are any of the following excluded in the protection equipment: lightning protective devices, interconnection of all grounds, and surge protective capacitors?

• Are there any surge protective capacitors that are not connected to each ungrounded service conductor?

corrosive conditions, is any insulation or sheath not approved for these conditions?

• Is there any Division II location where rigid metal conduit, electrical metallic tubing, or Types MI, MC, or ALS cable (with approved fittings) is not used?

• Are there any wireways, fittings, or boxes in which taps, joints, or terminal connections are made that are not provided with an effective means to prevent the escape of sparks or burning material?

• Where condensed vapors or liquids may collect on, or come into contact with the conductors or insulation, is the insulation not approved for this use?

GROUNDING CONNECTIONS

• Are there any locknut bushing or double-locknut types of contacts that are depended upon for bonding procedures? (NOTE: Bonding jumpers with proper fittings or other approved means must be used. This applies to all intervening raceways, fittings, boxes, and enclosures between hazardous areas and the point of grounding for service equipment.)

• If flexible conduit is used where permitted, is there any conduit without proper fittings for their bonding jumpers?

• Where there is an underground service conductor of a wiring system in a Class I location that is supplied from an overhead line, it is not protected by an approved lightning protective device?

• If a lightning protective device is used, is it connected to service conductors in any way except on the

• Is any capacitor unprotected by a 30-amp fuse of suitable type and voltage rating, or by an automatic circuit breaker of suitable type and rating?

• Is any capacitor not connected to the supply conductors on the supply side of the service disconnecting means?

FLEXIBLE CORDS

• Are there any flexible cords used for any purpose except as connections between a portable lamp (or any other fixed utilization equipment) and the fixed portion of the supply circuit?

• Are there any cords not approved for extra-hard usage?

• Is there any flexible cord that does not contain a grounding conductor in addition to the conductors in the circuit?

• Is there any flexible cord which is not connected to terminals or supply conductors in a manner consistant with the requirements?

• Is there any flexible cord supported by clamps (or other means) so that there is tension in the terminal connections?

• Is there any flexible cord which does not have a suitable seal where it enters into boxes, fittings, or enclosures of the explosion-proof type?

• Is there any cord that does not have a suitable dust/ignition-proof seal where it enters boxes, fittings, or enclosures of the Class II type?

/more/

Table 1-8. OSHA Check List—Wiring Systems (concluded)

CIRCUITS FOR SIGNALS, ALARMS, REMOTE CONTROLS, AND INTERCOMMUNICATION SYSTEMS

• Where accidental damage or breakdown of insulation might cause arcs, sparks, or high temperature, is any wiring method except one of the following used:

(a) Rigid metal conduit?

(b) Electrical metallic tubing?

(c) Type MI cable with approved termination fittings?

• Concerning the number of conductors in conduit or electrical metallic tubing, does the cross sectional area of all conductors exceed 40% of the area of the raceway?

• Are there any switches, circuit breakers, relays, contactors, fuses (which interrupt other than voice currents), current-breaking contacts for bells, horns, sirens, or other devices that produce sparks, arcs, not located in dust/ignition-proof enclosures? (NOTE: Heat-generating apparatus such as transformers, choke coils, and rectifiers must be enclosed, too.)

• If dust is electrically conducting or if it is the dust of hazardous metals, is any wiring used that is not approved for Class II locations?

• Are there any wiring or terminal connections of transformers or choke coils that are not provided with tight metal enclosures or without ventilating openings?

• Is there any switch, circuit breaker, or make/break contact of pushbuttons, relays, alarm bells, or horns that do not have an enclosure to suit the environment?

Start a violation inspection/abatement procedure on items that can be quickly and easily identified and repaired. Getting the plant physically clean is a good start. This means getting rid of debris in or around main feeder lines, clearing away extraneous items on the cable duct supports, and removing nonrelated materials that "just happen to be hanging" from the conduit.

WIRING ITEMS SURE TO BE CHECKED

Compliance with local building codes as well as with OSHA is a must. Check out these local or regional codes first, because they are usually more demanding than OSHA. But, if a case is found where the federal law is more stringent, obey that. If the laws disagree, OSHA takes precedence.

Main lines are usually the MC or Armor-Lock type. But when these are not used, incorporate cable ducts or metal troughs into the system. If these lines have to be unavoidably placed in damp or wet locations, use RHH or RHW cables for added protection. For installations in a corrosive atmosphere, the main lines can be coated with a protective jacket.

Grounding is a vital part of wiring safety inspection. OSHA is still picking up more violations here than anywhere else in the electrical area.

Make sure that electrical outlets have a proper service ground. Then make sure all your cord-connected tools and appliances are grounded (either by the three-wire method or by self-grounding).

It's recommended to use the portable power distribution systems for new construction areas or where extensive modernization/rearrangement is taking place. These "big boxes" allow technicians to safely plug in a number of cord-connected tools (power drills, saws) used on a major project. All receptacles are protected by their own circuit breakers and are grounded and weatherproof. An optional ground fault protective device is also available for added protection.

Any plant that uses DC power supplies should pay particular attention to that circuit. It's just as important to ground these systems as it is for the AC type, but they are often overlooked. (Cranes operated over combustible materials in Class III locations, however, do not need to be grounded.)

Check fixed equipment (including motor frames and craneways) to be sure they are grounded. It's easy to pass by a large metal structure such as a column that supports an overhead crane. If that ever became energized, it would present a great hazard.

Guard live parts in and around machinery. Some wiring may be exposed at this point so you must place the entire assembly in an approved cabinet, or put a permanent partition or wire enclosure around it. Warning signs should be placed around an installation of this sort, and it should be locked or sealed off if possible to prevent access by unqualified personnel.

Wires going into fixed machinery must be enclosed in conduit or placed within a metal shield. This prevents chafing and resulting failure of the insulation.

Shut-off devices can be mounted directly on this metal conduit, provided they are in a convenient location and the operator can readily see that it is a disconnect for the machine. To protect the operator, should he have to walk along or above a piece of equipment to check on a production process, shut-off devices must be located at intervals along the operator's route.

Panic shut-down devices are required for all your testing room facilities. These are red-glowing mushroom shaped lights that indicate when a test is in progress. An interlock prevents unauthorized entry.

Label all the service lines within cabinets or other enclosures. This quickly identifies the circuits and allows faster emergency repairs. Use moisture-resistant labels that won't peel off or deteriorate from the environment. It's not required by OSHA to maintain a log book to record dates of inspection, but this is very helpful in deciding when the time has come to update wiring or equipment.

Overcurrent devices must be a part of every wiring circuit. Put them in an approved enclosure and mount vertically. Be especially careful about utilizing the right enclosure if these devices are placed in damp or corrosive locations. Metal enclosures must be treated to combat the environment. Corrosion-resistant fiber-glass might be a good solution. The seals around the door and wiring, the wires and connection devices must also be able to prevent entrance of moisture.

CLEAR UP SLOPPY INSTALLATIONS

Any "temporary" wiring installation done will stick out like a sore thumb during an OSHA inspection.

A citation will probably occur for an open or broken junction box cover, a wiring connection made just by twisting wire together, or maybe an extension cord that runs across an aisle or under a door.

Some of the problems (and violations) involving the use of extension cords are a direct result of lack of power supply outlets. OSHA does not list a specific number that must be available but they can object if there are too few. The problem is easily spotted by snakes of extension cords winding around the floor. Occasionally, the cords are drawn through water, through a hole in the wall, or across an aisleway where a forklift truck can damage them, Locate and properly install more grounded receptacles where they will do the most good.

USE TEST INSTRUMENTS

First, give exposed wiring an "eyeball" examination to check insulation. Spot cracks, stripping away, or other damage. Then measure the suspect wire with an insulation tester.

Insulation testers are typically powered by a 9-volt battery, but for testing they convert to 500 volts. This abundant amount of voltage quickly shows up any flaw. The reading on the meter is then compared to the UL-standard listing for that particular wire.

It's beneficial to make a graph of the insulation resistance readings. When the graph takes a "quick dive" (indicating an insulation problem), soon there will be trouble. This graph idea is especially beneficial in checking out critical motor wiring and windings. Servicing can best be done on a planned maintenance schedule, rather than under emergency breakdown conditions.

Three-light ground testers are used to check for proper polarity and a valid ground. Simply plug into a wall socket.

Ground loop impedance testers plug into an outlet to give a direct resistance measurement.

Continuity testers measure the security of wiring connections and will show up a loose, hastily made hookup.

Field (or biomedical field) *probes* are hand held instruments that indicate by a "beep" sound where there is a static charge accumulation or current leakage. They point out exposed wiring, and can be used to check out a portable tool for proper ground by simply touching it.

2

Using the Language of the Electrical Engineer

To design the electrical portions of an industrial plant requires knowledge of power, lighting, and control. The viewpoint in the following chapters is that of an electrical engineer, designing a new industrial plant. This "role playing" experience will enable the reader to gain a better understanding of the elements that go into the design, to deal better with contractors and in-house designers, and to interpret drawings of an existing plant.

OBJECTIVES OF ELECTRICAL DESIGN

Three elements usually comprise the basis of electrical design; namely, technical proficiency, cost considerations, and overall schedules.

- *Technical Proficiency.* The electrical design should meet the plant's requirements, local and national codes and all safety requirements. The Occupational Safety and Health Act (OSHA) has incorporated most of the aspects of the National Electrical Code and is now the law of the land.

- *Cost Considerations.* Decisions relating to materials of construction, and first and operating costs should be analyzed using the principles of life-cycle costing. Electrical engineering and design manpower requirements should be established, monitored and controlled.

• *Schedules.* Schedules should be made for all engineering and construction activities. Delays in engineering or construction activities can be very costly. Monitoring progress, spotting areas of concern and implementing corrective action is required to ensure an orderly design.

ACTIVITIES OF THE ELECTRICAL ENGINEER/DESIGNER

Throughout this book you will be involved in the design of a hypothetical industrial plant. Since an industrial plant contains process, power generation, and office areas you will gain a broad exposure to electrical problems.

It will also be seen that to design the electrical portions of an industrial plant requires both an engineering and a design approach.

TYPICAL PROBLEM

The Ajax Company* is building a plant. The first step for the electrical engineer/designer is to help the client (plant) establish their needs. Many clients know what they want, but they need help in defining what has to be done. The design engineer studies all aspects of the client's requirements and establishes the design criteria. He must consider, for example:

• How to service the loads of the plant. This includes determining the voltage level to best service the load economically; should overhead or underground distribution be used; negotiations with the utility company to establish system requirements; and the type of reliability required.

• Type of lighting system required.

• Auxiliary systems required.

---

*The Ajax Company is fictitious, but the principles you will experience are not.

- Type of equipment required.
- Control considerations.
- Economic considerations.

Once the design criteria have been established the production of the job can begin. Design drawings and specifications to meet the job criteria can be made.

ENGINEERING ACTIVITIES

Using the above design criteria, the engineering activities include:
- Establishing criteria for One Line Diagram.
- Establishing budgets, schedules and manpower requirements.
- Writing specifications.
- Inspecting equipment.
- Coordinating activities of design, vendors, subcontractors and client.
- Checking vendor prints.
- Performing special studies.
- Preparing estimates.

DESIGN ACTIVITIES

Design activities are involved in the preparation of the various drawings required to convey installation information. Typical drawings include:

- *One Line or Single Line Diagram.* Figure 2-1 illustrates a typical single line diagram. This diagram is a simple schematic which identifies how power is distributed from the source to the user. Equipment such as switchgear, substations, motor control centers and motors are illustrated. The diagram also indicates the voltage levels, bus capacities, fuse or breaker ratings, key metering and relaying and other identification which will aid in describing the electrical distribution. Depending on the size of the system, sometimes several one line diagrams are needed. A main one line diagram illustrates the primary switchgear and substations. Motor Control Center one lines can then be used to show all motors and how they are being fed.

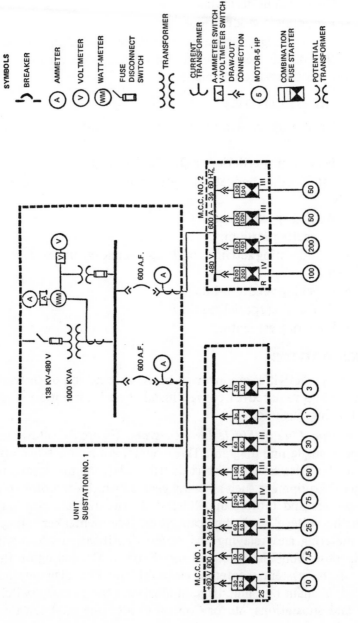

Figure 2-1. Typical One-Line Diagram

• *Power Plans.* Figure 2-2 illustrates a typical power plan. This diagram is a physical plan which is drawn to scale. It shows where all motors and loads are located and how they are fed. Conduit and cable sizes are indicated (if the project is large, conduit and cable sizes are indicated on separate sheets with only the description appearing on the power plan).

• *Elementary Diagrams.* Figure 2-3 illustrates a typical elementary diagram. This is a schematic which indicates how a system is controlled. Typical control devices such as pushbuttons, limit switches, level switches and pressure switches are used to energize relays, motor holding coils and solenoid valves. The elementary diagram indicates how a system operates, but not the physical properties of each element.

• *Interconnection Diagram.* A typical interconnection diagram is illustrated in Figure 2-4. The elementary diagram is used as the basis for this drawing. All relays are shown in their relative location. Terminal numbers and point-to-point wiring are shown. This drawing is used by the electrician to connect the wires to each terminal. Sometimes the information contained on the diagram is summarized on a schedule by a computer, thus eliminating the need for this drawing.

• *Lighting Plans.* A typical lighting plan is illustrated in Figure 2-5. This diagram is a physical plan which is drawn to scale. It shows the location of fixtures, outlets, and lamp circuiting. All lighting fixtures are either identified by a symbol on the lighting drawing or in a separate symbol list. Lighting panelboards may also be shown on this plan if space allows, or on separate schedule sheets. See Figure 2-6.

• *Details and Miscellaneous Diagrams.* Detail drawings may be used to supplement the power or lighting plans. These details may indicate a blow-up for field fabrication, a special support, an electrical room, outdoor substation layout, or any other item which needs further clarification.

Miscellaneous drawings may also be required to summarize materials, indicate instrument locations, or cover any drawing necessary to describe the complete electrical system.

• *Grounding Drawings.* To provide safety to personnel, a plant must be adequately grounded. A grounding drawing usual-

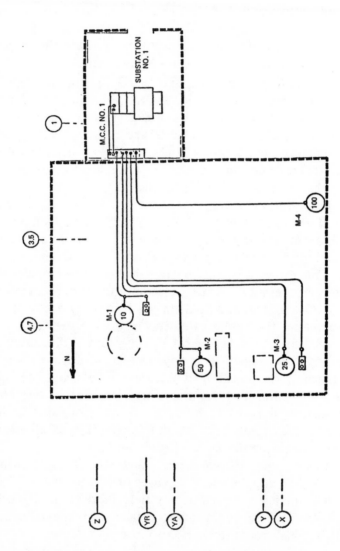

Figure 2-2. Typical Power Plan

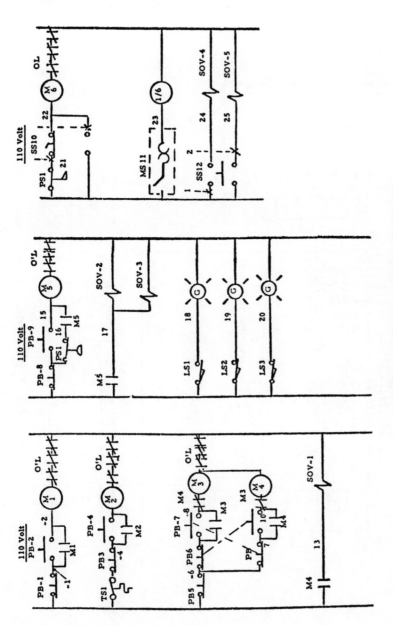

Figure 2-3. Typical Elementary Diagram

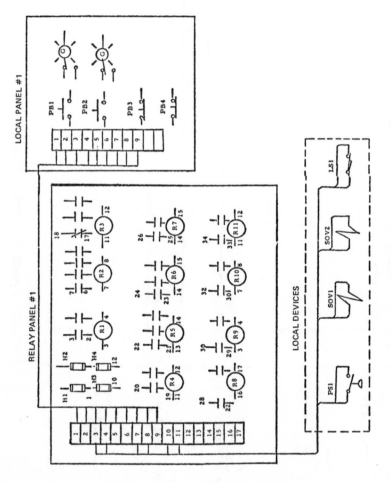

Figure 2-4. Typical Interconnection Diagram

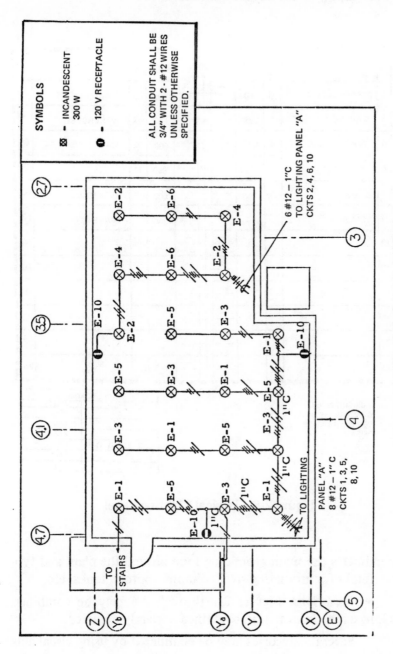

Figure 2-5. Typical Lighting Plan

LIGHTING PANEL "A"

CIRCUIT NO.	SERVICE	NO. OF OUTLETS	WATTS	NEUTRAL A B C	WATTS	NO. OF OUTLETS	SERVICE	CIRCUIT NO.
1	WAREHOUSE	5	1500		900	3	WAREHOUSE	2
3	WAREHOUSE	5	1500		600	2	WAREHOUSE	4
5	WAREHOUSE	4	1500		600	2	WAREHOUSE	6
7	H&V UNITS	-	320		990	11	WAREHOUSE STAIRS	8
9	BLANK	-	-		600	3	RECEPTACLE	10
11	BLANK		-		1200		SPARE	12
13								14
15								16
17								18
19								20
21								22
23								24

12 CIRCUIT PANEL
20 AMPERE BREAKER
120/208 V – 3 PHASE
4 WIRE

PANEL LOADING	
PHASE	WATTS
A	3710
B	2700
C	2100
CONNECTED LOAD	8510
SPARES	1200
TOTAL	9710

Figure 2-6. Typical Lighting Panel

ly indicates the main grounding loop around the plant and typical details for grounding steel columns, motors, tanks, etc.

• *Variations.* Figures 2-1 through 2-6 may be combined as long as the information contained is clearly depicted.

Several categories may be eliminated by using visual aids such as a model. Power plans which show motor runs can be eliminated if all conduit runs are shown on the model.

DESIGN ACTIVITY MANHOURS

When dealing with outside contractors and evaluating the scope of a project it is useful to have an understanding of how manhour estimates are made.

Table 2-1 summarizes the drawing types, drawings sizes, and the scales to which they are commonly drawn, plus the manhours required to produce the drawing. The hours for each drawing depend on the amount of detail shown and the type of firm that is making the drawing. For instance, an architectural firm designing a commercial lighting project may just choose

Table 2-1. Typical Drawing Requirements

DRAWING	SCALE	HOURS TO DESIGN DRAWING (including calculations)	TYPICAL DRAWING SIZE (Inches)
Lighting	1/8"=1'	30-50	30 x 42 24 x 36
Power	1/4"=1'	50-75	30 x 42 24 x 36
Lighting Schedules	None	10	9 x 12
Conduits & Cable Lists	None	10	12 x 18
Elementary Wiring	None	75-100	30 x 42 24 x 36
*One Lines	None	75-100	30 x 42 24 x 36
Interconnection	None	30-50	30 x 42 24 x 36
Grounding Drawings	1"=100' (Depends on Plot Plan)	30-40	30 x 42 24 x 36
Miscellaneous Details	Depends on Detail	50-100 Depends on Detail	30 x 42 24 x 36 9 x 12

*A unit substation and fifty motors can usually be fitted on a full-size one line diagram.

the fixture type and design a layout. A designer in an engineering firm would probably show in addition to the above, the circuits to each fixture and a lighting panelboard schedule. The more detail, the more hours. This table should serve as a guide. The practices in each firm and past job performances will be the prevalent factors in determining the drawing details and the estimated manhours.

In the following pages you will experience job situations. Each simulation experience will be denoted by SIM. The answer will be written below the problem. Cover the answer so that you can play the game.

SIM 2-1

Estimate the number of power and lighting layouts and approximate manhours for the following details.

Client: Ajax manufacturing plant "A"
          Basement — 50' x 200'
          Operating Floor 50' x 200'

Answer

From Table 2-1
Power Plan — 1/4"=1'
Thus 50' x 200' will fit on a 12" x 50" drawing.

Since a standard size drawing is 30" x 42", half of the operating and basement floor can fit on one drawing. Two drawings are required.

Lighting Plan — 1/8"=1'
$$\frac{50}{8} \times \frac{200}{8} = 6'' \times 25''$$

One drawing will be sufficient. 30" x 42".

SIM 2-2

Estimate the number of one line diagrams for the plant of SIM 2-1.

Given: 28 motors on two motor control centers with an estimated load of 800 KVA. Assume one substation will feed load.

Answer

Based on the above load, one unit substation and the associated Motor Control Centers will fit on one drawing.

SIM 2-3

Estimate the number of elementary and interconnection drawings for SIM 2-2. Each motor needs a separate stop-start control scheme. Control schemes should be provided so that 14 solenoid valves can be activated. All relays, pushbuttons, etc., are located on a local panel. Assume 100 elementary lines per drawing. Allow 2 spaces between each scheme

Answer

Estimating elementary and control schemes is a very difficult task. Usually at the beginning of the project it is difficult to get an exact description of the control.

Assume stop-start scheme. From Figure 2-3 two lines per scheme; thus 28 x 2 = 56 lines. Assume 2 spaces between each motor.

Scheme: 28 x 2 = 56 lines

Assume 2 lines per each solenoid scheme
                28 x 2 = 56 lines

Since more than 100 lines are required, two drawings are estimated.

In estimating the interconnection diagrams, assume an interconnection diagram is needed for each motor control center and local panel. By actually counting the number of terminals required, the number of drawings could be reduced later on.

SIM 2-4

Compile a drawing list with estimated manhours for problems SIM 2-1 through 2-3.

Answer

DRAWING NO.	DESCRIPTION	ESTIMATED MANHOURS
DWG 1	One Line Diagram	75
DWG 2	Power Plan	50
DWG 3	Power Plan	50
DWG 4	Lighting Diagram—Basement Operating	30
DWG 5	Grounding Drawing	30
DWG 6	Conduit and Cable Schedule Assume 2*	20
DWG 7	Lighting Schedule—Assume 2	20
DWG 8	Elementary Diagram	75
DWG 9	Elementary Diagram	75
DWG 10	Interconnection MCC No. 1	30
DWG 11	Interconnection MCC No. 2	30
DWG 12	Interconnection Local Devices	30
	TOTAL MANHOURS	515

*One drawing for Basement and one drawing for Operating Floor.

ENGINEERING ACTIVITY MANHOURS

It is more difficult to evaluate the engineering activities at the beginning of the project because many of the activities are involved with intangibles such as coordination.

From the details of the design an estimate can be made of the number of requisitions which are needed to purchase equipment.

• To write a specification using a previous one as a guide may take from 10 to 15 hours.

• To write a completely new specification may take 60 hours.

• Compiling material requisitions for quotation and purchase, evaluating sellers' quotes, and checking vendor prints could vary from 40 to 100 hours per requisition. The manhours required depends on the complexity of the equipment being purchased.

Many times engineering activities may be estimated as a percentage of design manhours. Coordination and general engineering activities may be from 10-20% for small projects below 5000 hours and 15-30% for larger ones which need more coordination. Specifications, requisitions and special studies should be added in separately.

SIM 2-5

Estimate the engineering hours for the design of 500 manhours as indicated in SIM 2-4. Assume no special studies; past specifications are available and minimum coordination time is required.

Answer

Substation Specification	10
Requisition and Vendor Prints	40
Motor Control Centers (2)	
Specifications	10
Requisition and Vendor Prints	40
Coordination—10% (500)	50
	150 manhours

Notice that the engineering activities on an industrial project are usually only a fraction of the total time required for design.

TYPES OF DESIGN ORGANIZATIONS

The design section of the facilities department of a plant or that of an outside consulting firm is usually organized either by departments or by tasks. The two organizations widely encountered are:

• *Department Oriented.* In the department oriented firm, each group (Electrical, Structural, Civil, Architectural, HVAC, Piping, and Plumbing) is separate from the other. The Department Head usually has his people grouped together. All information usually channels down from the Department Head.

• *Task Force.* Each Department Head chooses people for a project. These people move out of the department and operate as a team. Each individual usually gains responsibility since the departmental chain of command has disappeared.

In the "Project" or "Task Force" approach, a Project Engineer is assigned to the job to help coordinate the various disciplines.

KNOW THY VENDOR

Manufacturers provide many services in order to sell their products. Good relations with manufacturers can aid in electrical design.

Typical services include:

• Computer analysis for determining the most economical lighting system.

• Technical brochures.

• Cost information.

• Proven expertise in their field.

Since most of these services are free, good communications with vendors is extremely important. Remember vendors are offering these services as a means of selling their product. Be careful not to get "married" to one vendor, and to look objectively at all information he is offering.

Table 2-2 lists some of the vendors involved in specified areas. This list is neither complete nor a recommendation, but is presented merely as a guide to familiarize the reader with several commonly used vendors.

ELECTRICAL SCHEDULE

ALWAYS LAST

Electrical design cannot proceed without the motor horsepowers. For example, the heating and ventilating group must size their fans, the mechanical group must select their pumps, and the architectural group must select the automatic roll-up doors before all motor horsepower can be given to the electrical engineer. The electrical engineer is vulnerable to any changes from the other departments since their changes will probably affect the electric load.

Other aspects which affect design are firm equipment locations. This input is required to design the power and lighting drawings. A description of operation or a logic diagram is needed before the elementary and interconnection drawings can be designed. Thus the electrical design is usually the last to be finished on a project. Because of this, the electrical engineer is always under pressure to complete the design.

CRITICAL PATH

In many cases it is the electrical group which determines the critical path. Remember that the delivery of electrical equipment such as switchgear or high voltage bus duct may take up to a year to fabricate. Thus, one of the first activities the electrical engineer should do is purchase equipment. This means that many times an estimate of electrical loads must be used to purchase equipment in order to meet the schedules.

A typical schedule is illustrated in Figure 2-7.

JOB SIMULATION SUMMARY PROBLEM

JOB 1

Background

At the end of several chapters, a job simulation summary problem will be given. In these problems you will play the role of an Electrical Engineer working on Process Plant 2 of a grass roots (new) project for the Ajax Corporation. The plant is comprised of two identical modules. The response to each of these problems will be needed in order to complete the subsequent chapters.

Table 2-2. Vendor Check List

Vendor	Annunciators	Relays	P.B. & Lights	Motor Control Centers	Substations	H.V. Starters	Metering	Junct. Boxes	Transf. Sw.	Conduit	Distribution & Lgt. Panels	Conduit Fittings	Cable	Splicing Equip.	Term. Blks.	Timers	Closed Ckt. TV	Fire Alarm	Batteries	Light. Fixt.	Exit Light.	H.V. Fuses	L.V. Fuses	Terminal Boxes	Duct	Bus Duct	Fire Pump Controller	Engine Driven Fire Pump Cont.	Communication	Instrument Wiring	Pole Line
Panalarm (SCAM)	X																														
Russell & Stoll Co.	X																X														
Edwards Co.	X																	X													
Exide Ind. Div.																															
Square "D"		X	X	X							X			X	X																
Allen Bradley		X	X	X										X	X																
General Electric		X	X	X	X	X					X	X		X																	
Westinghouse		X	X	X	X	X					X			X																	
I.T.E.					X																										
ASCO								X																							
Anaconda												X																			
General Cable												X																			
Phelps Dodge												X																			
Motorola																X															
ADT																	X														
Autocall																	X														

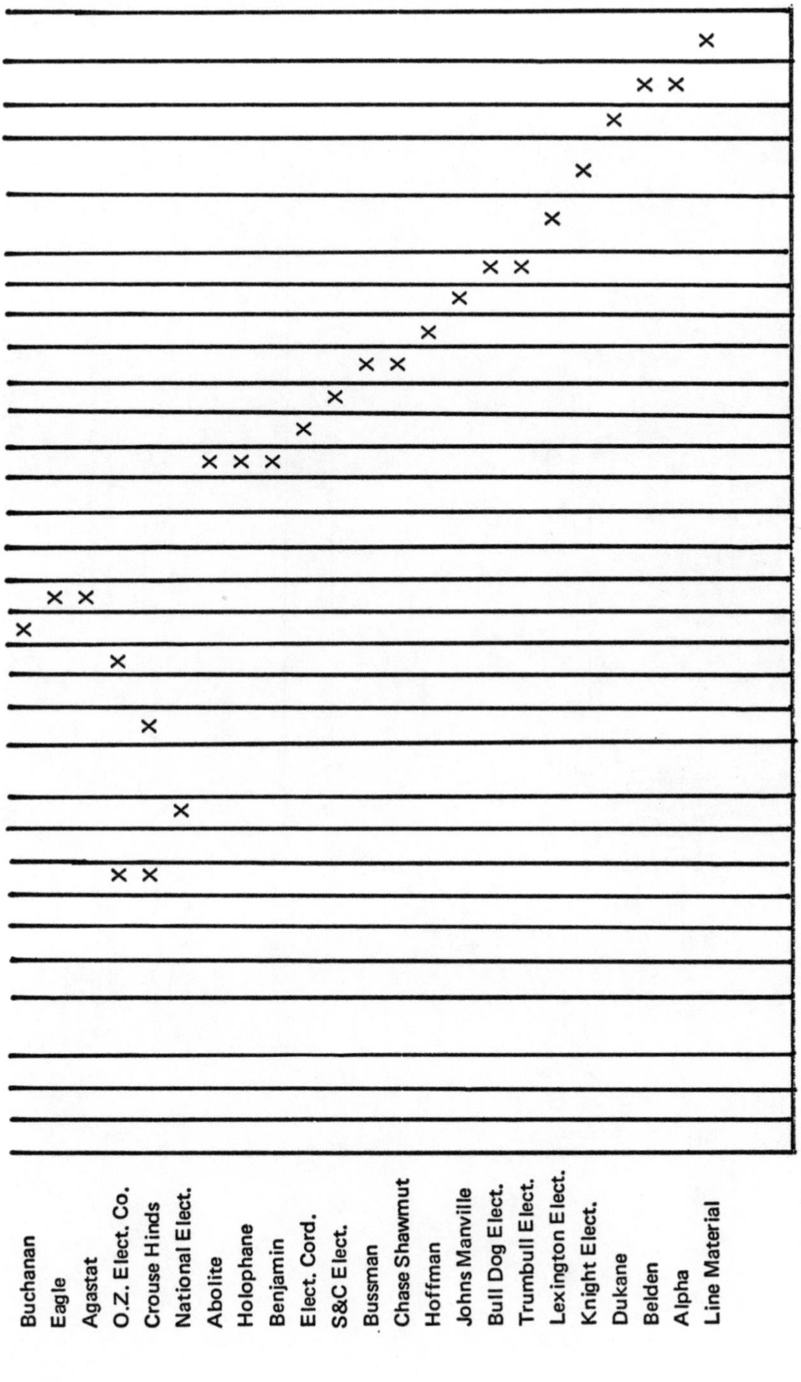

Buchanan
Eagle
Agastat
O.Z. Elect. Co.
Crouse Hinds
National Elect.
Abolite
Holophane
Benjamin
Elect. Cord.
S&C Elect.
Bussman
Chase Shawmut
Hoffman
Johns Manville
Bull Dog Elect.
Trumbull Elect.
Lexington Elect.
Knight Elect.
Dukane
Belden
Alpha
Line Material

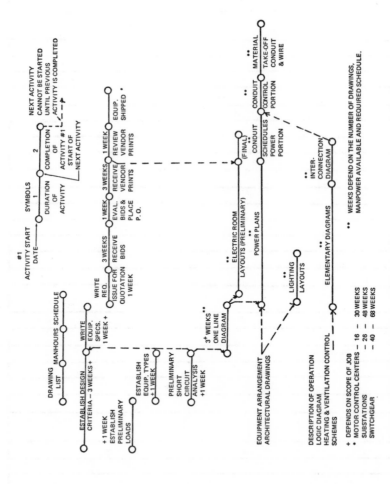

Figure 2-7. Schedule of Electrical Activities for Plant Design

The first task will be to identify the electrical loads from an equipment list which was given to the engineer by the plant. From this list indicate which loads require electric power and from which group you would expect to receive the information.

For example:

A — Architectural
M — Mechanical
H — Heating and Ventilating

Equipment List — Module #1

Note: Module #2 is identical.

D-1	Tank #1
AG-1	Agitator for Tank #1
H-2	Heat Exchanger
CF-3	Centrifuge
FP-4	Feed Pump
TP-5	Transfer Pump
CTP-6	Cooling Tower Feed Pump
D-7	Water Chest
V-8	Vessel #1
CT-9	Cooling Tower
HF-10	H&V Supply Fan
HF-11	H&V Exhaust Fan
UH-12	Unit Heater
BC-13	Brine Compressor
A-14	Air Operated Motor
T-15	Turbine #1
C-16	Conveyor
H-17	Hoist
ES-18	Exhaust Stack
SC-19	Self-cleaning Strainer
VS-20	Vacuum Separator
PO-21	Pneumatic Oscillator
RD-22	Roll-up Door

The plant has also indicated that one unit substation will handle the load and that each module should be fed from a motor control center.

The area of the Process Plant 2 is as follows:

Basement   20 ft x 400 ft
Operating Floor   20 ft x 400 ft

From the above information prepare a drawing list and submit the cost to do this project. Assume the project is complex (use higher hours per drawing) and that the unit rate is $30 per technical manhour which includes overhead, profit and fees.

Analysis

Once it is known which equipment requires a motor, it is necessary to obtain the electrical loads from each discipline and compile a motor list. A completed motor list for each module would look as follows:

EQUIP. NO.	NAME	HP	TYPE	DESIGN DISCIPLINE
AG-1	Agitator	60		M
CF-3	Centrifuge	100	Reversing	M
FP-4	Feed Pump	30		M
TP-5	Transfer Pump	10		M
CTP-6	Cooling Tower Feed Pump	25		M
CT-9	Cooling Tower	20	2-Speed	M
HF-10	H&V Supply Fan	40		H
HF-11	H&V Exhaust Fan	20		H
UH-12	Unit Heater	1/6		H
BC-13	Brine Compressor	50		M
C-16	Conveyor	20	Reversing	M
H-17	Hoist	5	Local starter by vendor	M
SC-19	Self-cleaning Strainer	3/4		M
RD-22	Roll-up Door	1/8	Local starter by vendor	A

Next, prepare a drawing list and estimate the manhours required to produce the drawings and specifications.

Typical Drawing List and Manpower Estimate

DESIGN		
DRAWING NO.	DESCRIPTION	ESTIMATED MANHOURS
DWG 1	One Line Diagram	100
DWG 2	Power Plan — Basement	75
DWG 3	Power Plan — Operating Floor	75
DWG 4	Lighting Diagram — Basement and Operating Floor	50
DWG 5	Grounding Drawing	40
DWG 6	Conduit & Cable Schedule (2)	20
DWG 7	Lighting Schedule (2)	20
DWG 8	Elementary Diagram	100
DWG 9	Elementary Diagram	100
DWG 10	Interconnection MCC No. 1 and Local Device	50
DWG 11	Interconnection MCC No. 2 and Local Device	50
		680 Hrs

ENGINEERING	
Substation Specification	15
Requisition and Vendor Prints	100
Motor Control Centers	15
Specifications, Requisitions, Vendor Prints	100
Engineering Coordination — 20% (680)	130
Total	360

TOTAL HOURS	1040 — say 1100 Hrs	
TOTAL COST	$33,000	

The $33,000 cost is probably on the high side since the two modules are identical and the time to do the second module would be less. Also the assumption that the job is complex does not seem reasonable, based on the scope just outlined.

As you can see, estimating is an "art" rather than an exact science.

3

Applying the
Fundamentals
of Power

The electrical engineer is faced with problems involved in establishing the total load of the plant, the power factor of the plant, and equipment selection. This chapter illustrates the basic power relationships. Through the power triangle and simple trigonometric relationships most power problems can be solved. A simplified table is included in this chapter to assist you in estimating the short-circuit current of the plant and selecting substation breakers.

THE POWER TRIANGLE

The total power requirement of a load is made up of two components: namely, the resistive part and the reactive part. The resistive portion of a load can not be added directly to the reactive component since it is essentially ninety degrees out of phase with the other. The pure resistive power is known as the watt, while the reactive power is referred to as the reactive volt amperes. To compute the total volt ampere load it is necessary to analyze the power triangle indicated below:

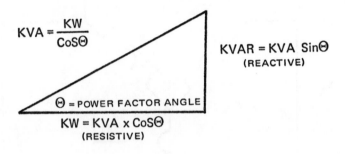

$$KVA = \frac{KW}{CoS\Theta}$$

$$KVAR = KVA \ Sin\Theta$$
(REACTIVE)

Θ = POWER FACTOR ANGLE

$$KW = KVA \times CoS\Theta$$
(RESISTIVE)

$K = 1000$

$W = $ Watts

$VA = $ Volt Amperes

$VAR = $ Volt Amperes Reactive

$\Theta = $ Angle Between KVA and KW

$CoS\Theta = $ Power Factor

$$\tan \Theta = \frac{KVAR}{KW}$$

RELATIONSHIPS

The windings of transformers and motors are usually connected in a wye or delta configuration. The relationships for line and phase voltages and currents are indicated below. The mathematical development of the $\sqrt{3}$ and the voltage and current relationships will not be discussed in this text.*

---

*For details see *Analyses of Electrical Circuits,* E. Brenner and M. Javid, McGraw-Hill, New York, 1959.

WYE CONNECTIONS

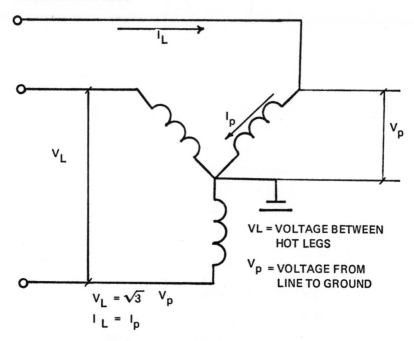

VL = VOLTAGE BETWEEN
      HOT LEGS

V_p = VOLTAGE FROM
      LINE TO GROUND

$$V_L = \sqrt{3} \ V_p$$
$$I_L = I_p$$

DELTA CONNECTIONS

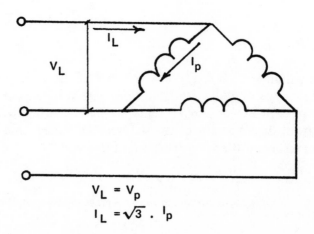

$$V_L = V_p$$
$$I_L = \sqrt{3} \cdot I_p$$

For a balanced 3-phase load

(Formula 3-1)   Power = $\underbrace{\sqrt{3}\ \ V_L\ \ I_L}\ $ CoSΘ

Watts	Volt	Power
	Amperes	Factor

For a balanced 1-phase load

(Formula 3-2) $P = V_L\ I_L\ $ CoSΘ

The primary windings of 13.8 Kv — 480 Volt unit substations are usually delta connected with the secondary wye connected.

MOTOR HORSEPOWER

The standard power rating of a motor is referred to as a horsepower. In order to relate the motor horsepower to a kilowatt (KW) multiply the horsepower by .746 (Conversion Factor) and divide by the motor efficiency.

(Formula 3-3) $KVA = \dfrac{HP \times .746}{\eta \times P.F.}$

HP = Motor Horsepower
η = Efficiency of Motor
$P.F.$ = Power Factor of Motor

Motor efficiencies and power factors vary with load. Typical values are shown in Table 3-1. Values are based on totally enclosed fan-cooled motors (TEFC) running at 1800 RPM "T" frame.

Table 3-1.

HP RANGE	3-30	40-100
η % at		
½ Load	83.3	89.2
¾ Load	85.8	90.7
Full Load	86.2	90.9
P.F. at		
½ Load	70.1	79.2
¾ Load	79.2	85.4
Full Load	83.5	87.4

IMPORTANCE OF POWER FACTOR

Transformer size is based on KVA. The closer Θ equals 0° or power factor approaches unity, the smaller the KVA. Many times utility companies have a power factor clause in their contract with the customer. The statement usually causes the customer to pay an additional power rate if the power factor of the plant deviates substantially from unity. The utility company wishes to maximize the efficiency of their transformers and associated equipment.

POWER FACTOR CORRECTION

The problem facing the electrical engineer is to determine the power factor of the plant and to install equipment such as capacitor banks or synchronous motors such that the overall power factor will meet the utility company's objectives.

Capacitor banks lower the total reactive KVAR by the value of the capacitors installed.

SIM 3-1

The secondary of a unit substation is a wye-connected transformer with neutral solidly grounded. The primary voltage is 13.8 KV and the secondary line voltage is 480 volts. Can a 277 volt lighting system be directly used?

Answer

Yes, since $V_L = 480$ and

$$V_p = V_L / \sqrt{3} = \frac{480}{\sqrt{3}} = 277 \text{ Volts}$$

SIM 3-2

A total motor horsepower load of 854 is made up of motors ranging from 40-100 horsepower. Calculate the connected KVA. Refer to Table 3-2 for trigonometric functions.

Table 3-2. Trigonometric Functions

Deg.	Sine	Tangent	Cotangent	Cosine	
0	.0000	.0000	.....	1.0000	90
1	.0175	.0175	57.29	.9998	89
2	.0349	.0349	28.636	.9994	88
3	.0523	.0524	19.081	.9986	87
4	.0698	.0699	14.301	.9976	86
5	.0872	.0875	11.430	.9962	85
6	.1045	.1051	9.5144	.9945	84
7	.1219	.1228	8.1443	.9925	83
8	.1392	.1405	7.1154	.9903	82
9	.1564	.1584	6.3138	.9877	81
10	.1736	.1763	5.6713	.9848	80
11	.1908	.1944	5.1446	.9816	79
12	.2079	.2126	4.7046	.9781	78
13	.2250	.2309	4.3315	.9744	77
14	.2419	.2493	4.0108	.9703	76
15	.2588	.2679	3.7321	.9659	75
16	.2756	.2867	3.4874	.9613	74
17	.2924	.3057	3.2709	.9563	73
18	.3090	.3249	3.0777	.9511	72
19	.3256	.3443	2.9042	.9455	71
20	.3420	.3640	2.7475	.9397	70
21	.3584	.3839	2.6051	.9336	69
22	.3746	.4040	2.4751	.9272	68
23	.3907	.4245	2.3559	.9205	67
24	.4067	.4452	2.2460	.9135	66
25	.4226	.4663	2.1445	.9063	65
26	.4384	.4877	2.0503	.8988	64
27	.4540	.5095	1.9626	.8910	63
28	.4695	.5317	1.8807	.8829	62
29	.4848	.5543	1.8040	.8746	61
30	.5000	.5774	1.7321	.8660	60
31	.5150	.6009	1.6643	.8572	59
32	.5299	.6249	1.6003	.8480	58
33	.5446	.6494	1.5399	.8387	57
34	.5592	.6745	1.4826	.8290	56
35	.5736	.7002	1.4281	.8192	55
36	.5878	.7265	1.3764	.8090	54
37	.6018	.7536	1.3270	.7986	53
38	.6157	.7813	1.2799	.7880	52
39	.6293	.8098	1.2349	.7771	51
40	.6428	.8391	1.1918	.7660	50
41	.6561	.8693	1.1504	.7547	49
42	.6691	.9004	1.1106	.7431	48
43	.6820	.9325	1.0724	.7314	47
44	.6947	.9657	1.0355	.7193	46
45	.7071	1.0000	1.0000	.7071	45
	Cosine	Cotangent	Tangent	Sine	Deg.

Answer

$$KVA = \frac{HP \times .746}{Motor\ Eff. \times Motor\ P.F.}$$

From Table 3-1 at full load

Eff. = .909 and P.F. = .87

$$KVA = \frac{854 \times .746}{.909 \times .87} = 806$$

SIM 3-3

As an initial approximation for sizing a transformer assume that a horsepower equals a KVA. Compare answer with SIM 3-2.

Answer

Total HP = 854

HP = KVA = 854

Problem answer = 806

For an initial estimate, equating horsepower to KVA is done in industry. This accuracy is in general good enough since motor horsepowers will probably change before the design is finished. The load at which the motor is operating is not established at the beginning of a project and this approach usually gives a conservative answer.

SIM 3-4

It is desired to operate the plant of SIM 3-2 at a power factor of .95. What approximate capacitor bank is required?

Answer

From SIM 3-2 the plant is operating at a power factor of .87. The power factor of .87 corresponds to an angle of 29°.

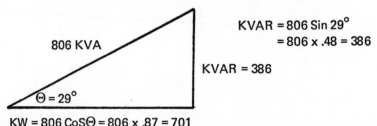

806 KVA

$\Theta = 29°$

KW = 806 CosΘ = 806 x .87 = 701

KVAR = 806 Sin 29°
= 806 x .48 = 386

KVAR = 386

A power factor of .95 is required.

$$Cos\Theta = .95$$
$$\Theta = 18°$$

The KVAR of 386 needs to be reduced by adding capacitors.

Remember KW does not change with different power factors, but KVA does.

Thus, the desired power triangle would look as follows:

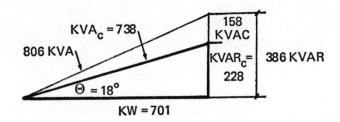

$$CoS\Theta = .95$$
$$\Theta = 18° \qquad Sin\ 18° = .31$$

$$KVA_c = \frac{701}{CoS\Theta} = \frac{701}{.95} = 738$$

Note: Power factor correction reduces total KVA
$$KVAR_c = 738\ Sin\ 18° = 738 \times .31 = 228$$
Capacitance Bank $= 386 - 228$
$$\cong 158\ KVAC$$

SIM 3-5

Assume that all motors in SIM 3-2 are not running at the same time. The diversity factor which takes into account the cycle time is assumed from previous plant experience to be 1.1. Indicate the minimum transformer size.

Answer

$$KVA_{Min.} = \frac{KVA_+}{Diversity\ Factor} = \frac{806}{1.1} = 732$$

Many times in industry the transformer capacity is simply based on the sum of the motor horsepowers plus an additional factor to take into account growth. The conservative sizing approach may not be too exact, but it does allow for normal changes in design and growth capacity. Remember electrical loads seldom shrink.

POWER FLOW CONCEPT

Power flowing is analogous to water flowing in a pipe. To supply several small water users, a large pipe services the plant at a high pressure. Several branches from the main pipe service various loads. Pressure reducing stations lower the main pressure to meet the requirements of each user. Similarly, a large feeder at a high voltage services a plant. Through switchgear breakers, the main feeder is distributed into smaller feeders. The switchgear breakers serve as a protector for each of the smaller feeders. Transformers are used to lower the voltage to the nominal value needed by the user.

ELECTRICAL EQUIPMENT

Electrical equipment commonly specified is as follows:

• *Switchgear-Breakers*—used to distribute power.

• *Unit Substation*—used to step down voltage. Consists of a high voltage disconnect switch, transformer and low-voltage breakers. Typical 480-volt transformer sizes are 300 KVA, 500 KVA, 750 KVA, 1000 KVA, 2000 KVA, 2500 KVA and 3000 KVA.

• *Motor Control Center (M.C.C.)*—a structure which houses starters and circuit breakers or fuses for motor control. It consists of the following:

(1) Thermal overload relays which guard against motor overloads;

(2) Fuse disconnect switches or breakers which protect the cable and motor and can be used as a disconnecting means;

(3) Contactors (relays) whose contacts are capable of opening and closing the power source to the motor.

MOTORS

• *Squirrel Cage Induction Motors* are commonly used. These motors require three power leads. For two-speed applications several different types of motors are available. Depending on the process requirements such as constant horsepower or constant torque, the windings of the motor are connected differently. The theory of two-speed operation is based on Formula 3-4.

(Formula 3-4)

$$\text{Frequency} = \frac{\text{No. of poles X speed}}{120}$$

Thus, if the frequency is fixed, the effective number of motor poles should be changed to change the speed. This can be accomplished by the manner in which the windings are connected. Two-speed motors require six power leads.

• *D. C. Motors* are used where speed control is essential. The speed of a D. C. Motor is changed by varying the field voltage through a rheostat. A D. C. Motor requires two power wires to the armature and two smaller cables for the field.

• *Synchronour Motors* are used when constant speed operation is essential. Synchronous motors are sometimes cheaper in the large horsepower categories when slow speed operation is required. Synchronous motors also are considered for power factor correction. A .8 P.F. synchronous motor will supply corrective KVARs to the system. A synchronous motor requires A. C. for power and D. C. for the field. Since many synchronous motors are self-excited, only the power cables are required to the motor.

MOTOR VOLTAGES

For fractional motors, 1/3 HP and below 115 volt, single phase is used. These motors are usually fed from a lighting panel and do not appear on the M.C.C. one-line diagram. A local starter consisting of a switch and over-load element is all that is usually required.

Motors 1/2 HP to 250 HP are usually fed from a 480-volt, 3-phase, motor control center or equivalent.

Motors 300 HP and above are usually fed at 2300 or 4160 volts. The reason for this is mostly economics, i.e., price of motor, starter, cable and transformer.

SIM 3-6

Determine the number of poles for a 3600 RPM motor—60 cycle service.

Answer

$$\text{No. of poles} = \frac{\text{Frequency} \times 120}{\text{Speed}} = \frac{60 \times 120}{3600} = 2$$

SIM 3-7

Next to each motor indicate the probable voltage rating and whether it is a single- or three-phase motor. Motor HP are: 1/4, 25, 150, 400.

Answer

MOTOR HP	VOLTAGE	NUMBER OF PHASES
1/4	110 V	1ϕ
25	460 V	3ϕ
150	460 V	3ϕ
500	2300 or 4.16 KV	3ϕ

SHORT-CIRCUIT CURRENTS

Faults occur for many reasons; deterioration of insulation, accidents, rats electrocuted across power leads, equipment failure and a multitude of other events. When a fault occurs a large short-circuit current flows. At first it has an initial peak or asymmetrical value, but after a period of time it will become symmetrical about the zero axis. Equipment must be rated to meet both the full-load currents and the short-circuit currents available.

RATING SUBSTATION BREAKERS (480V)

All breakers should be sized to meet the full-load current, the available short-circuit current, and must be able to coordinate with the system. Coordination of protective devices such as breakers means:

(a) That a protective device will not trip under normal operating conditions such as when the motor is started.

(b) That the protective element closest to the fault will open before the other devices upstream.

The impedance of the transformer limits the amount of short circuit which could flow. Table 3-3 illustrates how the transformer rating and utility system rating affect the short-circuit current. Use this table as a guide. A more detailed analysis is required on actual selection.

SIM 3-8

For a 1000 KVA transformer, 5.75% impedance, available primary short circuit of 250 MVA, determine the short-circuit current assuming 100% motor contribution.

Answer

From Table 3-3, the short-circuit current is 24,400 symmetrical amperes.

Table 3-3. Application Table: 480 Volts, Three Phase

Transformer Rating 3-Phase KVA and Impedance Percent	Maximum Short Circuit MVA Available from Primary System	Normal Load Continuous Current Amp	Short Circuit Current RMS Symmetrical Amp			Long-Time Instantaneous Recommended Min. Breaker Frame
			Transformer Alone	100% Motor Load	Combined	
1	2	3	4	5	6	7
300 5%	50 100 150 250 500 Unlimited	361	6500 6900 7000 7100 7200 7300	1400	7900 8300 8400 8500 8600 8700	225
500 5%	50 100 150 250 500 750 Unlimited	601	10000 10900 11300 11600 11800 12000	2400	12400 13300 13700 14000 14200 14200 14400	225
750 5.75%	50 100 150 250 500 750 Unlimited	902	12500 13900 14400 14900 15300 15400 15700	3600	16100 17500 18000 18500 18900 19000 19300	225

Amount / Rate	Col A	Term	Col B	Col C	Col D	Col E
1000 5.75%	1203	50	15500	4800	20300	225
		100	17800		22600	
		150	18800		23600	
		250	19600		24400	600
		500	20200		25000	
		750	20500		25300	
		Unlimited	20900		25700	
1500 5.75%	1804	50	20600	7200	27800	600
		100	24900		32100	
		150	26700		33900	
		250	28400		35600	1600
		500	29800		37000	
		750	30300		37500	
		Unlimited	31400		38600	
2000 5.75%	2406	50	24700	9600	34300	1600
		100	31100		40700	
		150	34000		43600	
		250	36700		46300	
		500	39100		48700	
		750	40000		49600	3000
		Unlimited	41900		51500	
2500 5.75%	3008	50	28000	12000	40000	1600
		100	36400		48400	
		150	40500		52500	3000
		250	44500		56500	
		500	48100		60100	
		750	49500		61500	
		Unlimited	52300		64300	
3000 5.75%	3607	50	30700	14400	45100	1600
		100	41200		55600	
		150	46500		60900	3000
		250	51900		66300	
		500	56800		71200	
		750	58700		73100	4000
		Unlimited	62700		77100	

[more]

Table 3-3. Application Table: 480 Volts, Three Phase (concluded)

Breaker Frame	Continuous Current Ratings Typical Trip Sizes	480V Breaker Rating	Short-Time Rating Amperes RMS Symmetrical
8	9	10	11
225	15, 20, 30, 40, 50, 70, 90, 100, 125, 150, 175, 200, 225	225	9,000
600	40, 50, 70, 90, 100, 125, 150, 175, 200, 225, 250, 300, 350, 400, 500, 600	600	22,000
1600	200, 225, 250, 275, 300, 350, 400, 500, 800, 1000, 1200, 1600	1600	50,000
3000	2000, 2500, 3000	3000	65,000
4000	2000, 2500, 3000, 4000	4000	85,000

Note: Usually at the beginning of a project if no utility data is available, assume unlimited short-circuit current and 100% motor load contribution.

SIM 3-9

A 1500 KVA substation secondary breaker feeds a 600 ampere Motor Control Center bus. Select a breaker frame to meet this load.

Answer

From Table 3-3, the 1600 ampere breaker is the minimum breaker size.

Based on a continuous current rating a 600 ampere frame breaker would have been sufficient, but the 600 ampere breaker can only handle a short-circuit current of 22,000 amperes. The 1600 ampere breaker can handle a short-circuit current up to 50,000 amperes and is good for 1600 continuous amperes. Thus, the frame size required is 1600 amperes. The trip rating which indicates when the breaker will open can be set at any value as indicated in Table 3-3. Assuming the breaker will co-ordinate with the load, a 600 ampere trip would be chosen.

JOB SIMULATION—SUMMARY PROBLEM

JOB 2

Background

(1) From the motor list established in Job 1 at the end of Chapter 2, indicate the rated voltage of each motor and if it is a single or three phase.

(2) The client wishes to know the power factor at which the plant is operating. Exclude motors below 3 HP from computations. Assume a lighting load of 40 KW. The plant is comprised of two identical modules (2 motors for each equipment number listed in Job 1). (Remember that KVAs at different power factors can not be added directly.)

(3)   Based on the total KVA of the plant determine the transformer size and the rating of each breaker. Assume an individual breaker feeds each module. The substation data will be sent to the three industries listed below for competitive bids:

Recommended vendors: ABC Industries
DEF Industries
GHI Industries

On the following pages are the responses to Job 2 and bids received, based on the correct answers.

Analysis

(1)   The client reviewed the motor list and expected to see the following:

Motor List — Module 1

MOTOR NO.	DESCRIPTION	HP	VOLTAGE	PHASE
AG-1	Agitator Motor	60	460	3
CF-3	Centrifuge Motor	100	460	3
FP-4	Feed Pump Motor	30	460	3
TP-5	Transfer Pump Motor	10	460	3
CTP-6	Cooling Tower Feed Pump Motor	25	460	3
CT-9	Cooling Tower Motor	20	460	3
HF-10	H&V Supply Fan Motor	40	460	3
HF-11	H&V Exhaust Fan Motor	20	460	3
UH-12	Unit Heater Motor	1/6	110	1
BC-13	Brine Compressor Motor	50	460	3
C-16	Conveyor Motor	20	460	3
H-17	Hoist Motor	5	460	3
SC-19	Self-Cleaning Strainer Motor	3/4	460	3
RD-22	Roll-Up Door Motor	1/8	110	1

(2)   Based on the motor list, the plant power factor was estimated at .87. Here's what the client expected to see:

Module #1

Lighting KW_3 = 40 Total

Motors 3-30	Motors 40-100
30	60
10	100
25	40
20	<u>50</u>
20	250
20	
<u>5</u>	
130	

At Full Load:

P.F. = 83.5	P.F. = 87.4
η = 86.2	η = 90.9

$$KVA_1 = \frac{130 \times .746}{.83 \times .86} = 135 \qquad KVA_2 = \frac{250 \times .746}{.90 \times .87} = 238$$

$$\begin{aligned} KW_1 &= KVA \; Cos\Theta \\ &= KVA \; .83 \\ &= 112 \end{aligned} \qquad \begin{aligned} KW_2 &= KVA_2 \; Cos\Theta \\ &= KVA \; .87 \\ &= 207 \end{aligned}$$

$$\begin{aligned} KVAR_1 &= KVA_1 Sin\Theta \\ \Theta &= 33° \end{aligned} \qquad \begin{aligned} KVAR_2 &= KVA_2 Sin\Theta \\ \Theta &= 29° \end{aligned}$$

$$\begin{aligned} KVAR_1 &= KVA_1 \times .54 \\ &= 135 \times .54 \\ &= 73 \end{aligned} \qquad \begin{aligned} KVAR_2 &= KVA_2 \times .48 \\ KVAR_2 &= 115 \end{aligned}$$

$$KW_{total} = KW_1 + KW_2 + KW_1 + KW_2 + KW_3 =$$

$$\qquad\qquad \underset{1}{Module} \quad \underset{2}{Module} \qquad = 678 \; KW$$

$$KVAR_{total} = KVAR_1 + KVAR_2 + KVAR_1 + KVAR_2$$
$$\text{Module 1} \qquad\qquad \text{Module 2}$$
$$= 73 + 115 + 73 + 115 = 376$$

$$KVA_{total} = \sqrt{(678)^2 \times (376)^2} = 774\ KVA$$

$$Co S\Theta = \frac{KW_{total}}{KVA_{total}} = \frac{678}{774} = .87$$

(3) The substation requisition should have included a 1000 KVA transformer and two 600 ampere feeder breakers.

$$\text{Transformer} = 1.25\ (\text{Total KVA}) = 1.25 \times 774$$
$$= 967\ KVA\ \text{size required.}$$

The closest transformer size is 1000 KVA.

The minimum breaker size based on Table 3-3 is 600 amperes. This size is required even though the total load on each breaker is $\frac{774}{2} = 387\ KVA$ or

$$\frac{387K}{\sqrt{3} \times 480} = 466\ \text{Amps}$$

The breaker must be sized to meet the short-circuit current of approximately 25,000 amps symmetrical.

Quotes received based on the substation requisition:

ABC Industries

Dear Sir:

In reply to subject inquiry, we are pleased to quote on the following:

1–480 Volt Substation including:

1. Incoming line compartment with load interrupter switch and current limiting fuse.
2. Dry type transformer 1000 KVA 13.8 KV to 480 volt wye.
3. Low voltage switchgear with main bus, feeder instrument and metering—2 breakers.

Equipment shall be arranged in accordance with layout attached.

Total Price is $24,000.

Shipment can be made 30 weeks after receipt of order.

Standard terms 30 days from date of invoice.

Very truly yours,

/s/ Al B. See

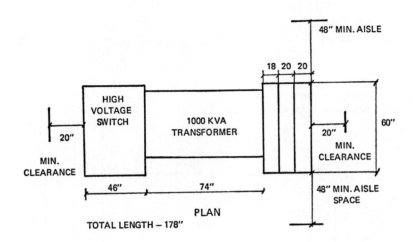

DEF Industries

Gentlemen:

We regret to inform you that at this time we can not quote. We appreciate your interest in DEF Industries and hope that we can be of service in the future.

Very truly yours,

/s/ Dee Frank

GHI Industries

Dear Sir:

We wish to offer our package completely in accordance with your specification.

The total price for the order is $24,500.

The above prices are quoted F.O.B. factory.

The earliest time equipment can be shipped is 40 weeks after receipt of order.

The attached layout shows the arrangement of the equipment.

Yours very truly,

/s/ Gee Hi Eye

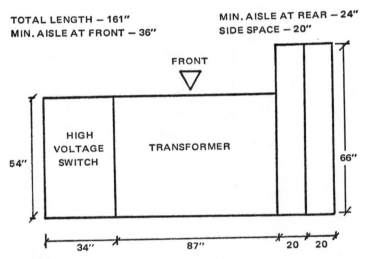

SUMMARY

Transformer ratings are based on KVA. Since loads constantly change, a sizing based on the sum of the motor and miscellaneous loads plus 1.25% gives a reasonable initial basis for determining capacity.

When determining the rating of equipment, always make sure that the equipment meets the short-circuit current of the plant.

4

Analyzing Power
Distribution Systems

The electrical engineer initiates one-line diagrams for new plants and interprets one-line diagrams of existing plants. This chapter illustrates some of the simple concepts involved in establishing a one-line diagram, determining system reliability, and designing a motor control center.

At the end of Chapter 3, three quotes were received from substation vendors. These quotes will be analyzed and a vendor will be recommended at the end of this chapter.

HOW TO DRAW A ONE-LINE DIAGRAM

An overall one-line diagram indicates where loads are located and how they are fed.

• The first step is to establish loads and their locations by communicating with the various engineers. (Remember initial design is based on your best estimate.)

• The next step is to determine the incoming voltage level based on available voltages from the utility company and the distribution voltage within the plant.

     (a)   For small plants to 10,000 KVA voltage levels may be 2300, 4160, 6900 or 13.8 KV.

(b)   For medium plants 10,000 KVA to 20,000 KVA voltage levels may be 13.8 KV.

(c)   For large plants above 20,000 KVA, 13.8 KV or 33 KV are typical values.

The advantages with the higher voltage levels are:

(a)   Feeders and feeder breakers can handle greater loads. (More economical at certain loads.)

(b)   For feeders which service distant loads, voltage drops are not as noticeable on the higher voltage system.

• The third step is to establish equipment types, sizes and ratings.

• The last step is to determine the system reliability required. The type of process and plant requirements are the deciding factors. The number of feeds and the number of transformers determine the degree of reliability of a system.

The three commonly used systems are the simple radial, primary selective, and secondary selective systems.

• *The Simple Radial System* is the most economical. As Figure 4-1 indicates, it is comprised of one feed and one transformer.

Figure 4-1
Simple Radial System

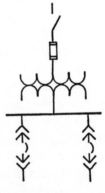

• *The Primary Selective System* is comprised of two feeds and two primary transformer disconnect switches. See Figure 4-2.

Figure 4-2
Primary Selective System

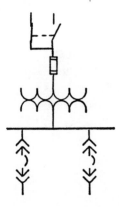

• *The Secondary Selective System* is the most reliable and the most expensive. As Figure 4-3 indicates, it is comprised of two complete substations joined by a tie breaker.

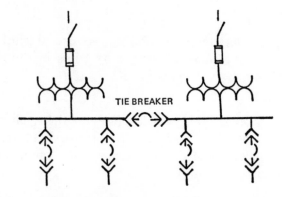

Figure 4-3. Secondary Selective System

SIM 4-1

Develop the one-line diagram for the Ajax Plant. The various steps for developing a one-line diagram are illustrated in SIM 4-1 through SIM 4-4.

Step 1: Establish loads and their locations.

	Approximate Load (KVA)
Administration Building	210
Machine Shop	340
Warehouse	185
Boiler House	675
Process Unit No. 1	890
Process Unit No. 2	765
	{ Use the load from Job 2b, Chapter 3

Locate loads on Figure 4-4.

Answer

See Figure 4-5.

SIM 4-2

Step 2: Determine distribution voltage to substations for SIM 4-1. Assume plant capacity may triple in the future and the utility primary voltage is 115 KV.

Answer

Total load is 3065 which represents the small plant category. Since the plant load may triple in the future, **13.8 KV** would be recommended since it meets the present load with future expansion.

SIM 4-3

Step 3: Establish equipment sizes for SIM 4-1. Substations should be located as close as possible to the load center. If loads are small (below 300 KVA) and are located near another load center, consideration should be given to combining it with other small loads.

Use standard size transformers illustrated in Table 3-3, Chapter 3, and size at least 1.25 times given load.

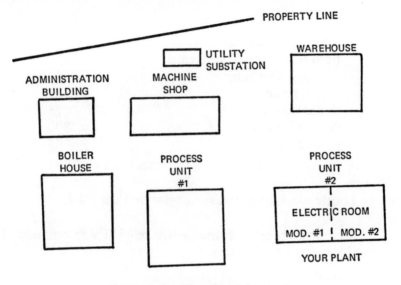

Figure 4-4. Ajax Plant Layout

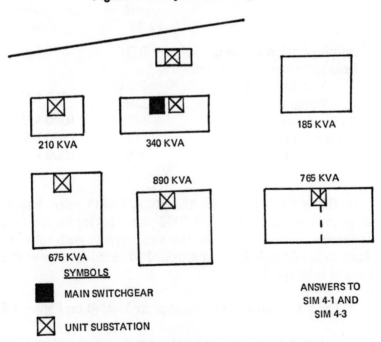

Figure 4-5. Answers to SIM 4-1 and SIM 4-3

	Load
Administration Building	210
Feed from same Transformer	
Machine Shop	340
Warehouse	185
Boiler House	675
Process Unit No. 1	890
Process Unit No. 2	_765_
Total Load	3065

Locate substations and switchgear on Figure 4-4.

Step 4: Comment on the system reliability recommended assuming a noncritical process.

    Answer

	Load *1.25*	*Transformer Size*
Administration Building	262	300 KVA
Feed from same Transformer		
Machine Shop	425 ⎫	
Warehouse	230 ⎭	750 KVA
Boiler House	845	1000 KVA
Process Unit No. 1	1112	1500 KVA
Process Unit No. 2	956	1000 KVA

The main feeder from the utility transformer should be sized to meet the total load of 3065 plus capacity for the future. It becomes impractical to run this size feeder to each substation. Thus the main switchgear is provided to distribute power to the various substations.

The substations and switchgear are located on Figure 4-5.

Unless the process is critical, a simple radial system is commonly used.

SIM 4-3

Draw a one-line diagram for SIM 4-1. Assume that up to 500 KVA can be put on each motor control center. Provide two switchgear breakers, one feeding the process substations and the second feeding the utility and auxiliary areas.

Answer   See Figure 4-6.

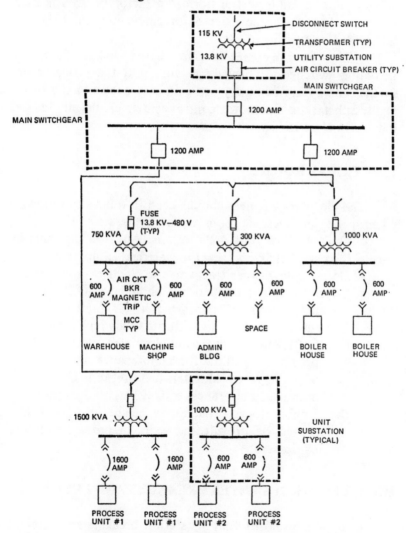

Figure 4-6. Overall One-Line Diagram

GENERAL TIPS FOR ERECTION

• Always size unit substations with growth capacity (25% growth capacity for transformers is common practice).
• A transformer with fans increases its rating. A 1000 KVA dry-type transformer with fans is good for 1333 KVA (33% increase). For an oil-type transformer a factor of 25% is used. Fans should only be considered for emergency conditions or for expanding existing plants.
• The question comes up as to where to locate equipment. Incoming switchgear is usually located near the property line so that the utility company can gain easy access to the equipment. Substations and motor control centers are usually located indoors in electrical rooms.

ELECTRICAL ROOMS

The electrical engineer should keep in mind the following when specifying electrical room requirements.
• Do not allow roof penetrations. Any roof opening increases the risk of fluid entering the electrical equipment.
• Do not allow other trades to use electrical room space.
• General ambient temperature should be 40°C. (Special equipment, such as computers, may require air conditioning.)
• Lay out electrical rooms with the following in mind.
   (a) Sufficient aisle space and door clearances should be provided to allow for maintenance and replacing of damaged transformers and breakers. Use the recommended clearances established by the vendor.
   (b) Double doors of adequate height (usually 8 feet) should be provided at exits in order to remove equipment.

MOTOR CONTROL CENTER BREAKERS AND FUSES

It should be noted that substation breakers are different from switchgear breakers and motor control center breakers.

Switchgear breakers may be of the "vacuum type," whereas substation breakers may be of the "magnetic air circuit type" and motor control centers may be of the "molded case" type.

Table 4-1 summarizes breaker and starter sizes for motor control centers. Table 4-2 summarizes dual element fuse and switch sizes for motor control centers. The fuse and breaker sizes indicated in these tables are based on vendor's data. As long as the values are below specified values listed in the National Electrical Code and coordinate with the motor, the selection is satisfactory.

There are several types of fuses commonly used. Each type is characterized by its time to isolate the fault, interrupting rating, and current limiting property. Fuse types include: standard fuses, time delay fuses, current limiting fuses, and dual element fuses.

ADVANTAGES OF FUSES OVER BREAKERS

The advantages of fuses over breakers are:

• Higher interrupting ratings (100,000 amps-dual element fuses).

• Current limiting action—will limit the short current downstream of fuse.

• Lower cost.

• Less affected by corrosive atmosphere.

• Less affected by moisture.

• Less affected by dust.

ADVANTAGES OF BREAKERS OVER FUSES

The advantages of breakers over fuses are:
• Resetable.
• Electrically operated breakers can be remotely operated.
• Adjustable characteristics.

Table 4-1. Combination Breaker-Starter

MOTOR HP	STARTER SIZE	BREAKER TRIP*	BREAKER FRAME**	M.C.C. SPACE†	BREAKER TYPE	TRIP RANGE	ASYM. AMPS
1	I	15	100	14"	FA	15-100	15,000
1½	I	15	100	14"	JA	70-225	20,000
2	I	15	100	14"	KA	70-225	25,000
3	I	15	100	14"	LA	125-400	35,000
5	I	15	100	14"			
7½	I	30	100	14"			
10	I	40	100	14"			
15	I	50	100	14"			
20	II	50	100	14"			
25	II	50	100	14"			
30	II	70	100	14"			
40	III	100	100	28"			
50	III	100	100	28"			
60	III	125	225	28"			
75	IV	150	225	28"			
100	IV	200	225	28"			
125	V	225	225	42"			
150	V	300	400	42"			
200	V	350	400	42"			

* Check chosen vendor for specific recommendations.

** Minimum size — check short-circuit rating.

† Based on M.C.C. Vendor's data for FVNR (Full Voltage Non-Reversing Starters). Check chosen vendor for specific details.

Table 4-2. Combination Fuse-Starter

MOTOR HP	460 V F.L.A.	FUSE*	SWITCH	M.C.C.** SPACE
1	1.8	4	30	14"
1½	2.6	5	30	14"
2	3.4	8	30	14"
3	4.8	10	30	14"
5	7.6	15	30	14"
7½	11	20	30	14"
10	14	25	30	14"
15	21	30	30	14"
20	27	40	60	14"
25	34	50	60	14"
30	40	60	60	28"
40	52	80	100	28"
50	65	100	100	28"
60	77	125	200	42"
75	96	150	200	42"
100	124	200	200	42"
125	156	250	400	70"
150	180	300	400	70"
200	240	400	400	70"

*Based on Dual Element Fuses.

**Based on M.C.C. Vendor's data for FVNR Starters.
Check chosen vendor for specific details.

SIM 4-4

Indicate the starter size for 10, 30 and 100 HP motors.

Answer (from Table 4-1)

10 HP	Size I
30 HP	Size III
100 HP	Size IV

SIM 4-5

The short-circuit current available at a motor control center is 15,000 amperes asymmetrical. Indicate the frame (continuous rating) and trip (current at which breaker will open) sizes for 7½, 30, 60 and 100 HP motors.

Answer (from Table 4-1)

Horsepower	Trip Size	Frame Size
7½	30	100
30	70	100
60	125	225
100	200	225

SIM 4-6

The short-circuit current available at a motor control center is 25,000 amperes asymmetrical.

Repeat SIM 4-5.

Answer (from Table 4-1)

Horsepower	Trip Size	Frame Size	
7½	70	225	Min. Size
30	70	225	Breaker
60	125	225	KA–25,000
100	200	225	

Note: For large short-circuit currents, breakers are impractical as indicated above. Either the short-circuit current should be decreased or fuses should be used instead of breakers.

SIM 4-7

Indicate the fuse and switch sizes for the following: 3, 10, 50, 75 HP motors.

Answer (from Table 4-2)

Horsepower	Fuse Size	Switch Size
3	10	30
10	25	30
50	100	100
75	150	200

MOTOR CONTROL CENTER LAYOUTS

Figure 4-7 shows a typical outline for a M.C.C. Dimensions vary between vendors and the space allocated for wiring depends on whether cables enter from top or bottom and if terminal blocks are required in upper or lower section.

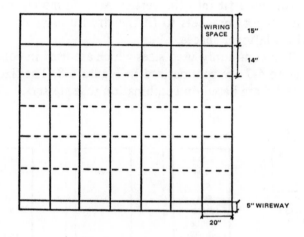

Figure 4-7. M.C.C. Outline
(Consult Vendor for Specific Dimensions)

Usually terminal blocks are located in the individual starter cubicle and the top or bottom is used only for wiring between sections.

Referring to Figure 4-7, a vertical section is a structure with an overall height of 90", a width of 20" and a depth of nominally 13"-20". It includes a horizontal feeder bus at the top and a vertical bus bar to accept the plug-in motor and starter units. A unit is the motor starter-disconnect module that fits into the vertical section and is covered by a door. Thus, the motor control center consists of vertical sections bolted together and connected with a common horizontal feeder bus bar. The working height is the available space in any section for motor control center units.

LAYOUT

When designing a motor control center keep in mind the following:

- Larger units should be placed near the bottom of a section for easier maintenance.
- Place the various required units in as many sections as necessary to accommodate them.
- All units fit into the vertical section merely by moving the unit support brackets to fit the structure. Filler plates can be used for leftover spaces.
- The commonly used sizes which are used in conjunction with Figure 4-7 are summarized in Figure 4-8. The sizes shown in Figure 4-8 are based on combination fuse starters.

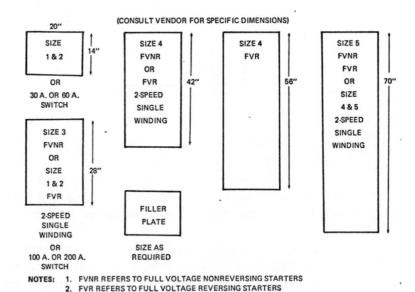

NOTES: 1. FVNR REFERS TO FULL VOLTAGE NONREVERSING STARTERS
2. FVR REFERS TO FULL VOLTAGE REVERSING STARTERS

Figure 4-8.
(Consult Vendor for Specific Dimensions)

Note: The minimum bus for a motor control center is 600 amperes. Initial sizing is usually based on 400 to 500 HP per motor control center. This is usually adequate. A detailed load check should be made when the design is firm.

M.C.C. ONE-LINE DIAGRAM

Typical symbols used for a M.C.C. one-line diagram are illustrated in Figure 4-9.

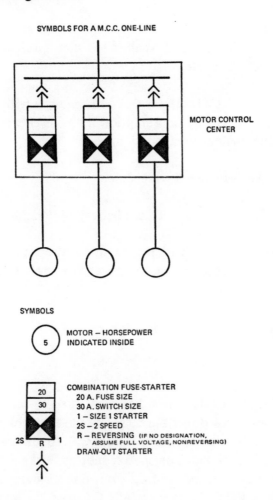

Figure 4-9. Symbols for a M.C.C. One-Line Diagram

ANALYZING BIDS

It is the responsibility of the electrical engineer to recommend vendors to build the electrical equipment for the plant. To get competitive bids, requests for quotations are initiated with several bidders. After the bids have been received, they must be evaluated in terms of quality, costs and schedule in particular.

• Check to see if vendor has met specifications. Check capacities, rating, etc.

• Look at advantages and disadvantages with each vendor. Physical size, energy consumption, noise level, weight, etc. may affect the final recommendation.

• Analyze delivery schedule.

• Evaluate cost picture. Check if all items are included.

Sometimes a client may wish to direct purchase an item due to overriding factors, such as spare parts or to match an existing installation.

JOB SIMULATION—SUMMARY PROBLEM

JOB 3a

The electrical engineer has been asked by the client to issue a requisition for direct purchase for the motor control centers. Assume a 30 Amp switch in each M.C.C. to take care of lighting loads and fractional horsepower motors. Make a sketch of the proposed layout and a M.C.C. one-line diagram. Assume each module is fed from a separate M.C.C. Use the horsepower and data of Job 2a, Chapter 3. Typical forms are illustrated.

Forms for Job 3a

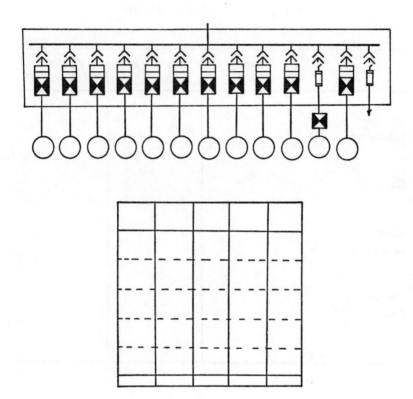

JOB 3b

Based on the substation quotes received in Job 2c, Chapter 3, recommend a vendor. Required delivery date is 35 weeks from date of order.

Complete.

Simplified Bid Analysis Form

	Vendor's Name	*Vendor's Name*	*Vendor's Name*
Cost per Substation			
Delivery			
Meets Specification			
Overall Area			

Bidder
Recommended

Reason:

JOB 3ç

Based on the substation and M.C.C. layout, estimate the area required for the electrical equipment room. Include two Motor Control Centers and one substation in the room.

Analysis

JOB 3a:

A typical response the client would expect to Job 3a would look as follows:

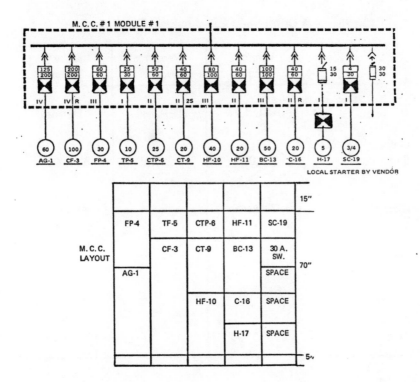

The quote from the M.C.C. vendor, based on the requisition is as follows:

Dear Sir:

The total cost for the Motor Control Center you outlined in your inquiry is $8,000 per Motor Control Center. For two Motor Control Centers the total cost is $16,000.

Delivery can be made in 16 weeks after drawing approval.

Very truly yours,

/s/ Marvin C. Cole

JOB 3b: A Completed Bid Analysis Form

	Vendor's Name	Vendor's Name	Vendor's Name
	ABC Ind.	DEF Ind.	GHI Ind.
Cost per Substation	$24,000	Declined	$24,500
Delivery	30 weeks		40 weeks
Meets Specification	Yes		Yes
Overall Area	178" X 60"		161" X 66"

Bidder Recommended   ABC   Industries

Reason: Only vendor which meets required delivery.
                    Lowest price.

JOB 3c: A Possible Electric Room Layout

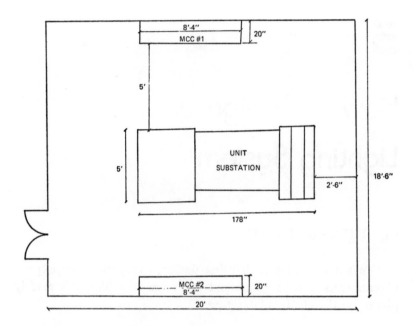

SUMMARY

The one-line diagram is the single most important drawing for the electrical design. It serves as the basis for:

Short-circuit calculations

Coordination studies

Power factor correction

Equipment selection

Power plans

In addition, the one-line diagram is a must for plant maintenance. By use of the one-line diagram, problem areas can be located and corrective action can be taken.

5

How to Design a Lighting System

LET THERE BE LIGHT!

When it comes to lighting design, everyone is an expert. If you walk into a plant and can't see, the blame is readily placed on the electrical designer.

LIGHTING DEFINITIONS

Footcandle − A measure of illumination−one standard candle power measured one foot away.
Lumen − A measure of lamp intensity.

CHOOSING LIGHTING LEVELS

The electrical engineer can determine the recommended levels of illumination by referring to the Illuminating Engineering Society Handbook RP-15 on Levels of Illumination, American National Standards Institute A11.1−1973, June 1973, and FEA Guidelines, Table 5-1.

Table 5-1. Recommended Maximum Lighting Levels

TASK OR AREA	FOOTCANDLE LEVELS	HOW MEASURED
Hallways or corridors	10 ± 5	Measured average, minimum 1 footcandle.
Work and circulation areas surrounding work stations	30 ± 5	Measured average.
Normal office work, such as reading and writing (on task only), store shelves, and general display areas	50 ± 10	Measured at work station.
Prolonged office work which is somewhat difficult visually (on task only)	75 ± 15	Measured at work station.
Prolonged office work which is visually difficult and critical in nature (on task only)	100 ± 20	Measured at work station.
Industrial tasks	ANSI-A11.1-1973	As maximum.

Many clients may wish to use levels consistent with their old plants, but it is up to the engineer to recommend lighting levels consistent with present design practices.

Another factor influencing levels of illumination is energy conservation. In an effort to conserve electricity, Table 5-1 may be used for "at the task" illumination levels. Aisles between task areas are maintained at one-third of the "at the task" levels. A balance between energy conservation lighting levels and previous practice levels should be made.

CHOOSING A LIGHTING SYSTEM

The major question facing the electrical engineer is what type of lighting system to use. He may be guided by past practice of the plant, but this is not the best technique. The type of lighting system to use depends on the physical characteristics of the room and the first and operating cost for the lighting system. Since many vendors have computer programs available to determine the best fixture type, the engineer should not overlook this time saving source.

In general the lamp with the lowest first cost (incandescent) has the shortest life and poorest lumen output per watt. This means that even though the first cost of the incandescent system is less than a high-intensity discharge system (mercury vapor), the operating cost of the incandescent system will be much higher than the other.

Table 5-2 summarizes commonly used high-intensity discharge (HID) lamps. From this table it is seen that the high-pressure sodium (HPS) lamp has a very high lumen output per watt. Other factors influence final selection such as the time to restart and the color characteristics of the lamp. Another characteristic of lamps is that lumen depreciation occurs with hours of usage. This point is also shown in Table 5-2.

SHORT CUTS FOR ESTIMATING LIGHTING LOADS

One of the first jobs the engineer should do is to estimate the lighting load. This estimate is needed for the one-line diagrams and for the HVAC heat load analysis. If the lighting can be designed, actual KW loads can be computed. Otherwise the following values may be useful:

• Direct or semi-direct mercury vapor varies linearly from 3 to 5.5 watts per square foot for footcandle levels of 50 and 100 respectively.

• Direct or semi-direct incandescent varies linearly from 5.5 to 11 watts per square foot for footcandle levels of 50 and 100 respectively.

• Direct or semi-direct fluorescent varies linearly from 1.9 to 3.7 watts per square foot for footcandle levels of 50 and 100 respectively.

As previously indicated, different lamp types affect lamp intensity so the above is at best a very rough estimate. But it may be all that is needed to determine order-of-magnitudes for lighting transformers, etc.

Table 5-2. High Intensity Discharge Lamp Characteristics
(Reprinted with permission from *Electrical Consultant*)

PER CENT INITIAL LUMENS

(Mfr. or ANSI No.)		RATED * LUMENS	BURNING HRS. OF OPERATION IN 1000'S								
			2	4	6	8	10	12	16	20	24
LAMPS—METAL HALIDE											
G.E. Multivapor											
400w	MV-400 BUH	34,000	87	80	76	72	70				
1000w	MV-1000 BUH	100,000	87	80	75	71	69				
G.E. I-Line											
400w	MV400/BU/I	34,000	13	86	80	75	70	66			
1000w	MV1000/BU/I	88,000	87	80	75	72	70				
Sylvania Metalarc											
175w	M175/BU	14,000	83	76	72	71					
250w	M250/BU-HOR	20,500	87	79	74	69					
400w	M400/BU-HOR	34,000	88	81	76	72	69	67			
1000w	M1000/BU-HOR	100,000	88	80	75	.72	71				
Sylvania Super-Metalarc											
175w	M175/HOR	15,000	83	76	72	71					
400w	M400/HOR	40,000	88	81	76	72	69	67			
Westinghouse—Metal Halide											
400w	MH400/BU/4	34,000	78	72	69	68	68	67			
1000w	MH1000/BU	100,000	86	78	73	71	69				
LAMPS—MERCURY VAPOR											
(DX) Deluxe White (Typical all Mfrs.)											
175w	H39KC-175/DX	8,600	94	92	89	86	83	81	76	70	65
250w	H37KC-250/DX	13,000	95	92	89	86	83	81	76	70	65
400w	H33GL-400/DX	23,000	93	91	88	85	83	79	71	67	60
1000w	H36GW-1000/DX	63,000	92	87	82	77	72	68	61	55	50
G.E. Warm DX											
175w	H39-KC-175/WDX	6,500	95	91	88	85	81	77	72	66	60
250w	H37KC-250/WDX	9,500	93	89	85	80	76	72	63	55	47
400w	H33GL-400/WDX	20,000	93	90	86	82	78	74	67	59	51
1000w	H36GW-1000/WDX	58,000				68		60			
Westinghouse "Beauty Lite"											
175w	H39-KC-175/R	8,500	93	88	84	80	76	73	66	60	53
250w	H37KC-250/R	13,000	91	84	79	75	73	70	66	60	52
400w	H33GL-400/R	23,000	93	87	83	79	76	73	67	61	56
1000w	H36GW-1000/R	63,000	89	83	78	73	69	66	61	57	54
Westinghouse "Style-Tone"											
175w	H39KC-175/N	7,000	93	88	84	80	76	73	66	60	53
250w	H37KC-250/N	11,000	91	84	79	75	73	70	66	60	52
~400w	H33GL-400/N	19,500	93	87	83	79	76	73	67	61	56
LAMPS—HPS											
G.E. "Lucalox"											
100w	LU100/BU	9,500	97	94	90	85	80	75			
150w	LU150/BU	16,000	97	94	90	85	80	75			
250w	LU250/BU .	25,500	99	97	94	91	87	83			
250w	LU250/BU/S	30,000	99	97	94	91	87	83			
400w	LU400	50,000	98	95	94	90	88	85	80	71	
1000w	LU1000	140,000	97	94	92	90	87	83			
Sylvania "Lumalux-2"											
250w	LU250BU	25,500	98	96	92	89	85	81			
400w	LU400BU	50,000	98	96	92	89	85	81			

* Lamp data subject to change and should be verified by latest lamp manufacturer's data.

/more/

Table 5-2. High Intensity Discharge Lamp Characteristics (concluded)

PER CENT INITIAL LUMENS

(Mfr. or ANSI No.)		RATED · LUMENS	BURNING HRS. OF OPERATION IN 1000'S								
			2	4	6	8	10	12	16	20	24
Westinghouse "Ceramalux"											
150w	C150	16,000	99	97	95	91	88	84			
250w	C250	25,500	99	97	95	91	88	84			
400w	C400	50,000	99	97	95	91	88	84			
1000w	C1000BU	130,000	97	93	90	87	84				
Norelco											
250w	LU250	25,500	97	95	93	91	88	84			
400w	LU400	47,000	97	95	93	91	88	84			
Sylvania "Unalux"											
150w	ULX150	12,000	98	95	92	89	85	80			
360w	ULX360	36,000	98	96	94	92	90	88			
G.E. "E-Z Lux"											
150w		13,000	consult lamp manufacturer								

Lamp data subject to change and should be verified by latest lamp manufacturer's data.

SIM 5-1

For the office area of Process Plant No. 1 recommend a desired footcandle level.

Answer

From Table 5-1, the desired footcandle level is 50.

SIM 5-2

For direct mercury vapor lighting, estimate the lighting load, using a footcandle level of 50.

Basement	50' X 200'
Operating Floor	50' X 200'

Answer

A 50 F.C. level requires 3 watts per square foot of lighting.

Total Area (50 X 200)2 = 20,000 sq. ft.

Total Load = 3 X 20 X 10^3 = 60 KW

LIGHTING METHODS

The zonal cavity method has been the conventional way to calculate the number of luminaires required. This method is desirable when a uniformly maintained lighting level is required throughout an area. With today's emphasis on efficient lighting design it is necessary to pay more attention to the factors which determine lighting levels. The zonal cavity method is based on Formula 5-1.

(Formula 5-1)

$$\text{Footcandle Maintained} = \frac{\text{No. of fixtures X lamps/fixture X lumens/lamp X C.U. X LLF}}{\text{Area}}$$

C.U. = Coefficient of utilization

LLF = Light loss factor

In order to reduce the number of lamps which in turn reduces energy consumption it is necessary to use efficient lamps, reduce the light loss factor, and insure a good coefficient of utilization. The factors which comprise Formula 5-1, namely the coefficient of utilization and light loss factor, are important to lighting and are discussed in detail.

COEFFICIENT OF UTILIZATION

The coefficient of utilization makes allowances for light absorbed or reflected by walls and the ceiling, and the fixture itself. It represents the ratio of the lumens reaching the working plane to the total lumens generated by the lamp. Table 5-3 illustrates the form in which a vendor summarized the data used for determining the coefficient of utilization.

To determine the coefficient of utilization the room cavity ratio, wall reflectance, and effective ceiling cavity reflectance must be known.

Most data assumes a 20% effective floor cavity reflectance. To determine the coefficient of utilization:

(a) Estimate wall and ceiling reflectances.
Typical values are shown in Table 5-4.

Table 5-3. Vendor Data for 175 Watt Mercury Vapor Lamp—Medium Spread Deflector
Coefficients of Utilization/Effective Floor Cavity Reflectance 20% (pFC)

% REFLECTANCE EFF. CEIL. (pCC)	WALL (pW)	ROOM CAVITY RATIO									
		1	2	3	4	5	6	7	8	9	10
80	50	0.854	0.779	0.711	0.647	0.591	0.539	0.490	0.446	0.407	0.355
	30	0.828	0.739	0.664	0.594	0.533	0.481	0.432	0.388	0.349	0.296
	10	0.805	0.705	0.626	0.552	0.491	0.440	0.392	0.347	0.309	0.258
70	50	0.832	0.761	0.698	0.635	0.578	0.530	0.483	0.438	0.401	0.349
	30	0.808	0.724	0.653	0.585	0.526	0.475	0.426	0.384	0.345	0.295
	10	0.786	0.695	0.618	0.546	0.486	0.434	0.387	0.344	0.308	0.256
50	50	0.788	0.725	0.669	0.610	0.558	0.511	0.466	0.424	0.388	0.339
	30	0.770	0.696	0.632	0.568	0.513	0.464	0.416	0.375	0.338	0.288
	10	0.754	0.670	0.602	0.534	0.478	0.428	0.382	0.339	0.303	0.253
30	50	0.750	0.694	0.642	0.587	0.539	0.495	0.450	0.412	0.377	0.329
	30	0.736	0.671	0.612	0.552	0.499	0.453	0.408	0.367	0.331	0.282
	10	0.722	0.649	0.586	0.523	0.469	0.421	0.375	0.335	0.299	0.249
10	50	0.716	0.665	0.618	0.566	0.521	0.479	0.438	0.399	0.366	0.319
	30	0.704	0.645	0.592	0.536	0.487	0.442	0.399	0.360	0.325	0.276
	10	0.693	0.628	0.571	0.511	0.460	0.413	0.370	0.330	0.294	0.245

Table 5-4. Typical Reflection Factors

COLOR	REFLECTION FACTOR
White and very light tints	.75
Medium blue-green, yellow or gray	.50
Dark gray, medium blue	.30
Dark blue, brown, dark green, and wood finishes	.10

   (b) Determine room cavity ratio.
        Refer to Figure 5-1 and Table 5-5.
   (c) Determine effective ceiling cavity reflectance (pCC).
        Refer to Figure 5-1, Tables 5-5 and 5-6.

Step (a)

To calculate the coefficient of utilization an estimate must be made of the reflection factors for walls and ceiling. Typical values are shown in Table 5-4.

Steps (b) and (c)

Once the wall and ceiling reflectances are estimated it is necessary to analyze the room configuration to determine the effective reflectances. Any room is made up of a series of cavities which have effective reflectances with respect to each other and the work plane. Figure 5-1 indicates the basic cavities.

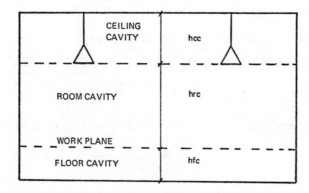

Figure 5-1.

Table 5-5. Cavity Ratios

Room Dimensions		Ceiling or Room Cavity Depth																			
Width	Length	1.0	1.5	2.0	2.5	3.0	3.5	4.0	5.0	6.0	7.0	8	9	10	11	12	14	16	20	25	30
8	8	1.2	1.9	2.5	3.1	3.8	4.4	5.0	6.2	7.5	8.8	10.0	11.2	12.5	—	—	—	—	—	—	—
	10	1.1	1.7	2.2	2.8	3.4	3.9	4.5	5.6	6.8	7.9	9.0	10.1	11.2	12.4	—	—	—	—	—	—
	14	1.0	1.5	2.0	2.5	2.9	3.4	3.9	4.9	5.9	6.9	7.9	8.8	9.8	10.8	11.8	—	—	—	—	—
	20	0.9	1.3	1.8	2.2	2.6	3.1	3.5	4.4	5.2	6.1	7.0	7.9	8.8	9.6	10.5	12.2	—	—	—	—
	30	0.8	1.2	1.6	2.0	2.4	2.8	3.2	4.0	4.8	5.5	6.3	7.1	7.9	8.7	9.5	11.1	12.7	—	—	—
	40	0.8	1.1	1.5	1.9	2.2	2.6	3.0	3.8	4.5	5.2	6.0	6.8	7.5	8.2	9.0	10.5	12.0	—	—	—
10	10	1.0	1.5	2.0	2.5	3.0	3.5	4.0	5.0	6.0	7.0	8.0	9.0	10.0	11.0	12.0	—	—	—	—	—
	14	0.9	1.3	1.7	2.1	2.6	3.0	3.4	4.3	5.1	6.0	6.9	7.7	8.6	9.4	10.3	12.0	—	—	—	—
	20	0.8	1.1	1.5	1.9	2.2	2.6	3.0	3.8	4.5	5.2	6.0	6.8	7.5	8.2	9.0	10.5	12.0	—	—	—
	30	0.7	1.0	1.3	1.7	2.0	2.3	2.7	3.3	4.0	4.7	5.3	6.0	6.7	7.3	8.0	9.3	10.7	—	—	—
	40	0.6	0.9	1.2	1.6	1.9	2.2	2.5	3.1	3.8	4.4	5.0	5.6	6.2	6.9	7.5	8.8	10.0	12.5	—	—
	60	0.6	0.9	1.2	1.5	1.8	2.0	2.3	2.9	3.5	4.1	4.7	5.2	5.8	6.4	7.0	8.2	9.3	11.7	—	—
12	12	0.8	1.2	1.7	2.1	2.5	2.9	3.3	4.2	5.0	5.8	6.7	7.5	8.3	9.2	10.0	11.7	—	—	—	—
	16	0.7	1.1	1.5	1.8	2.2	2.6	2.9	3.6	4.4	5.1	5.8	6.6	7.3	8.0	8.8	10.2	11.7	—	—	—
	24	0.6	0.9	1.2	1.6	1.9	2.2	2.5	3.1	3.8	4.4	5.0	5.6	6.2	6.9	7.5	8.8	10.0	12.5	—	—
	36	0.6	0.8	1.1	1.4	1.7	1.9	2.2	2.8	3.3	3.9	4.4	5.0	5.6	6.1	6.7	7.8	8.9	11.1	—	—
	50	0.5	0.8	1.0	1.3	1.6	1.8	2.1	2.6	3.1	3.6	4.1	4.6	5.2	5.7	6.2	7.2	8.3	10.3	12.9	—
	70	0.5	0.7	1.0	1.2	1.5	1.7	2.0	2.4	2.9	3.4	3.9	4.4	4.9	5.4	5.9	6.8	7.8	9.8	12.2	—
14	14	0.7	1.1	1.4	1.8	2.1	2.5	2.9	3.6	4.3	5.0	5.7	6.4	7.1	7.9	8.6	10.0	11.4	—	—	—
	20	0.6	0.9	1.2	1.5	1.8	2.1	2.4	3.0	3.6	4.2	4.9	5.5	6.1	6.7	7.3	8.5	9.7	12.1	—	—
	30	0.5	0.8	1.0	1.3	1.6	1.8	2.1	2.6	3.1	3.7	4.2	4.7	5.2	5.8	6.3	7.3	8.4	10.5	—	—
	42	0.5	0.7	1.0	1.2	1.4	1.7	1.9	2.4	2.9	3.3	3.8	4.3	4.8	5.2	5.7	6.7	7.6	9.5	11.9	—
	60	0.4	0.7	0.9	1.1	1.3	1.5	1.8	2.2	2.6	3.1	3.5	4.0	4.4	4.8	5.3	6.2	7.0	8.8	11.0	—
	90	0.4	0.6	0.8	1.0	1.2	1.4	1.7	2.1	2.5	2.9	3.3	3.7	4.1	4.5	5.0	5.8	6.6	8.3	10.3	12.4
17	17	0.6	0.9	1.2	1.5	1.8	2.1	2.4	2.9	3.5	4.1	4.7	5.3	5.9	6.5	7.1	8.2	9.4	11.8	—	—
	25	0.5	0.7	1.0	1.2	1.5	1.7	2.0	2.5	3.0	3.5	4.0	4.4	4.9	5.4	5.9	6.9	7.9	9.9	12.4	—
	35	0.4	0.7	0.9	1.1	1.3	1.5	1.7	2.2	2.6	3.1	3.5	3.9	4.4	4.8	5.2	6.1	7.0	8.7	10.9	—
	50	0.4	0.6	0.8	1.0	1.2	1.4	1.6	2.0	2.4	2.8	3.2	3.5	3.9	4.3	4.7	5.5	6.3	7.9	9.9	11.8
	80	0.4	0.5	0.7	0.9	1.1	1.2	1.4	1.8	2.1	2.5	2.9	3.2	3.6	3.9	4.3	5.0	5.7	7.1	8.9	10.7
	120	0.3	0.5	0.7	0.8	1.0	1.2	1.3	1.7	2.0	2.4	2.7	3.0	3.4	3.7	4.0	4.7	5.4	6.7	8.4	10.1
20	20	0.5	0.8	1.0	1.2	1.5	1.8	2.0	2.5	3.0	3.5	4.0	4.5	5.0	5.5	6.0	7.0	8.0	10.0	12.5	—
	30	0.4	0.6	0.8	1.0	1.2	1.5	1.7	2.1	2.5	2.9	3.3	3.8	4.2	4.6	5.0	5.8	6.7	8.3	10.4	12.5
	45	0.4	0.5	0.7	0.9	1.1	1.3	1.4	1.8	2.2	2.5	2.9	3.2	3.6	4.0	4.3	5.1	5.8	7.2	9.0	10.8
	60	0.3	0.5	0.7	0.8	1.0	1.2	1.3	1.7	2.0	2.3	2.7	3.0	3.3	3.7	4.0	4.7	5.3	6.7	8.3	10.0
	90	0.3	0.5	0.6	0.8	0.9	1.1	1.2	1.5	1.8	2.1	2.4	2.8	3.1	3.4	3.7	4.3	4.9	6.1	7.6	9.2
	150	0.3	0.4	0.6	0.7	0.8	1.0	1.1	1.4	1.7	2.0	2.3	2.6	2.8	3.1	3.4	4.0	4.5	5.7	7.1	8.5
24	24	0.4	0.6	0.8	1.0	1.2	1.5	1.7	2.1	2.5	2.9	3.3	3.8	4.2	4.6	5.0	5.8	6.7	8.3	10.4	12.5
	32	0.4	0.5	0.7	0.9	1.1	1.3	1.5	1.8	2.2	2.6	2.9	3.3	3.6	4.0	4.4	5.1	5.8	7.3	9.1	10.9
	50	0.3	0.5	0.6	0.8	0.9	1.1	1.2	1.5	1.8	2.2	2.5	2.8	3.1	3.4	3.7	4.3	4.9	6.2	7.7	9.2
	70	0.3	0.4	0.6	0.7	0.8	1.0	1.1	1.4	1.7	2.0	2.2	2.5	2.8	3.1	3.4	3.9	4.5	5.6	7.0	8.4
	100	0.3	0.4	0.5	0.6	0.8	0.9	1.0	1.3	1.6	1.8	2.1	2.3	2.6	2.8	3.1	3.6	4.1	5.2	6.5	7.8
	160	0.2	0.4	0.5	0.6	0.7	0.8	1.0	1.2	1.4	1.7	1.9	2.2	2.4	2.6	2.9	3.4	3.8	4.8	6.0	7.2

		C30	C36	C42	C50	C60	C75	C100	C150	C200	C300	C500
30	30 45 60 90 150 200	10.0 8.2 7.4 6.7 5.9 5.6	8.3 7.2 6.7 5.7 5.2 4.9	7.1 6.0 5.2 4.6 4.3 4.2	6.0 5.1 4.5 4.0 3.5	5.0 4.5 3.5 3.0	4.0 3.3 2.7 2.5	3.0 2.2 2.0	2.5 2.0	1.5 1.2	0.8	0.6
36	36 50 75 100 150 200	8.4 6.6 6.0 5.0 4.7	6.9 6.5 5.1 4.3 4.1	5.9 5.0 4.4 3.6 3.5	5.0 4.3 3.7 3.0 2.9	4.2 3.9 2.9 2.5	3.3 2.7 2.3 2.1	2.5 1.9 1.7	1.7 1.2	1.2 1.0	0.7	0.5
42	42 60 90 140 200 300	6.7 5.5 5.0 4.5 3.7	5.6 4.8 4.1 3.3 3.3	4.7 3.5 3.1 2.9 2.8	4.0 3.0 2.7 2.3	3.3 2.7 2.3 2.0	2.7 2.1 1.8 1.7	2.0 1.5 1.3	1.3 1.0	1.0 0.8	0.6	0.4
50	50 70 100 150 300	5.4 4.0 3.6 3.2 3.0	4.4 3.8 3.0 2.8 2.6	3.8 3.2 2.8 2.5 2.2	3.2 2.7 2.1 1.9	2.7 2.1 1.6	2.1 1.5 1.3	1.6 1.1	1.1	0.8 0.7	0.5	0.3
60	60 100 150 300	4.7 3.5 3.1 2.8 2.6	3.9 3.3 2.9 2.4 2.3	3.3 2.8 2.4 2.2 1.9	2.8 2.4 1.9 1.6	2.3 1.9 1.4	1.9 1.5 1.2	1.4 0.9	0.9 0.7	0.7 0.6	0.5	0.3
75	75 120 200 300	4.0 3.3 3.0 2.4 2.2	3.3 2.9 2.3 2.1 2.0	2.8 2.4 1.9 1.7 1.7	2.4 2.0 1.6 1.4	2.0 1.6 1.2	1.6 1.3 1.0	1.2 0.8	0.8 0.6	0.6 0.5	0.4	0.2
100	100 200 300	3.7 3.0 2.5 2.2 2.0	3.0 2.3 2.1 1.9 1.8	2.6 2.1 1.7 1.6	2.2 1.9 1.5 1.3	1.8 1.5 1.1	1.5 1.0 0.9	1.1 0.8 0.7	0.7 0.6	0.6 0.5	0.4	0.2
150	150 300	3.3 2.5 2.2 2.0 1.9	2.8 2.0 1.9 1.6	2.4 2.0 1.5 1.4	2.0 1.5 1.3 1.1	1.7 1.3 1.0	1.3 1.1 0.9	1.0 0.7 0.7	0.7 0.5	0.5 0.4	0.3	0.2
200	200 300	3.0 2.2 2.0 1.8 1.7	2.5 1.8 1.7 1.5	2.1 1.8 1.4 1.3	1.8 1.5 1.2 1.0	1.5 1.2 0.9	1.2 1.0 0.7	0.9 0.6 0.5	0.5 0.4	0.5 0.4	0.3	0.2
300	300	2.7 2.2 1.8 1.6 1.5	2.2 1.7 1.5 1.3 1.3	1.9 1.4 1.2 1.1	1.6 1.2 1.1 0.9	1.3 1.1 0.8	1.0 0.7 0.7	0.8 0.6 0.5	0.5 0.4	0.4 0.3	0.3	0.2
500	500	2.3 1.9 1.6 1.4 1.3	1.9 1.7 1.3 1.2 1.1	1.6 1.4 1.1 1.0 0.9	1.4 1.2 0.9 0.8	1.2 0.9 0.8 0.7	0.9 0.8 0.6	0.7 0.5 0.4	0.5 0.3	0.3 0.3	0.2	0.1

Table 5-6. Per Cent Effective Ceiling or Floor Cavity Reflectance for Various Reflectance Combinations

% Ceiling or Floor Reflectance	10			30				50			70			80				90			
% Wall Reflectance	10	30	50	10	30	50	65	30	50	70	30	50	70	30	50	70	80	30	50	70	90
0	10	10	10	30	30	30	30	50	50	50	70	70	70	80	80	80	80	90	90	90	90
0.1	10	10	10	29	29	30	30	48	49	50	68	69	69	78	78	79	79	87	88	89	90
0.2	10	10	10	28	29	29	30	47	48	49	66	67	68	76	77	78	79	85	86	88	89
0.3	9	10	10	27	28	29	30	46	47	49	64	66	68	74	75	77	78	83	85	87	89
0.4	9	10	11	26	27	29	30	45	46	48	63	65	67	72	74	76	78	81	83	86	88
0.5	9	10	11	25	27	28	29	44	46	48	61	64	66	70	73	75	77	78	81	85	88
0.6	9	10	11	25	26	28	29	43	45	47	59	62	65	68	71	75	77	76	80	84	88
0.7	8	10	11	24	26	28	29	42	44	47	58	61	65	66	70	74	76	74	78	83	88
0.8	8	10	11	23	25	27	29	41	43	47	56	60	64	65	69	73	75	73	77	82	87
0.9	8	9	11	22	25	27	29	40	43	46	55	59	63	63	68	72	75	71	76	81	87
1.0	8	9	11	22	24	27	29	39	42	46	53	58	63	61	66	71	74	69	74	80	86
1.1	8	9	11	21	24	26	29	38	41	46	52	57	62	60	65	71	74	67	73	79	86
1.2	7	9	12	20	23	26	29	37	41	45	50	56	61	58	64	70	73	65	72	78	86
1.3	7	9	12	20	23	26	29	36	40	45	49	55	61	57	63	69	73	64	70	78	85
1.4	7	9	12	19	22	26	28	35	40	45	48	54	60	55	62	68	72	62	69	77	85
1.5	7	9	12	18	22	25	28	34	39	44	47	53	59	54	61	68	72	61	68	76	85
1.6	7	9	12	18	21	25	28	33	39	44	45	52	59	53	60	67	71	59	66	75	85
1.7	7	9	12	17	21	25	28	32	38	44	44	51	58	52	59	66	71	58	65	74	84
1.8	6	9	12	17	21	25	28	32	37	43	43	50	57	50	58	65	70	56	64	73	84
1.9	6	9	12	16	20	25	28	31	37	43	42	49	57	49	57	65	70	55	63	73	84
2.0	6	9	12	16	20	24	28	30	37	43	41	48	56	48	56	64	69	53	62	72	83

Ceiling or Floor Cavity Ratio

2.1	2.2	2.3	2.4	2.5	2.6	2.7	2.8	2.9	3.0	3.1	3.2	3.3	3.4	3.5	3.6	3.7	3.8	3.9	4.0	4.1	4.2	4.3	4.4	4.5	4.6	4.7	4.8	4.9	5.0
6	6	6	6	6	5	5	5	5	5	5	5	5	5	5	5	4	4	4	4	4	4	4	4	4	4	4	4	4	4
9	9	9	9	9	9	9	9	9	8	8	8	8	8	8	8	8	8	8	8	8	8	8	8	8	8	8	8	8	8
13	13	13	13	13	13	13	13	13	13	13	13	13	13	13	13	13	13	13	14	13	13	13	13	14	14	14	14	14	14
16	15	15	14	14	13	13	13	12	12	12	11	11	11	11	10	10	10	10	9	9	9	9	8	8	8	8	8	7	7
20	19	19	19	18	18	18	18	17	17	17	16	16	16	16	15	15	15	15	15	14	14	14	14	14	14	13	13	13	13
24	24	24	24	23	23	23	23	23	22	22	22	22	22	22	21	21	21	21	21	21	20	20	20	20	20	20	19	19	19
28	28	28	28	27	27	27	27	27	27	27	27	27	27	26	26	26	26	26	26	26	26	26	26	25	25	25	25	25	25
29	29	28	27	27	26	26	25	25	24	24	23	23	22	22	21	21	21	20	20	20	19	19	19	19	18	18	18	18	17
36	36	35	35	34	34	33	33	33	32	32	31	31	31	30	30	30	29	29	29	28	28	28	27	27	27	26	26	26	26
43	42	42	42	41	41	41	41	40	40	40	40	39	39	39	39	38	38	38	38	37	37	37	37	37	36	36	36	36	36
40	39	38	37	36	35	34	33	33	32	31	30	30	29	29	28	27	27	26	26	25	25	25	24	24	24	23	23	23	22
47	46	46	45	44	43	43	42	41	40	40	39	39	38	38	37	37	36	36	35	35	34	34	34	33	33	33	32	32	32
56	55	54	54	53	53	52	52	51	51	50	50	49	49	48	48	48	47	47	46	46	46	45	45	45	44	44	44	43	43
47	45	44	43	42	41	40	39	38	38	37	36	35	34	33	33	32	31	30	30	29	29	28	27		26	26	25	25	25
55	54	53	52	51	50	49	48	48	47	46	45	44	44	43	42	42	41	40	40	39	39	38	38	37	37	36	36	35	35
63	63	62	61	61	60	60	59	58	58	57	57	56	56	55	54	54	53	53	52	52	51	51	51	50	50	49	49	49	48
69	68	68	67	67	66	66	66	65	65	64	64	64	63	63	62	62	62	61	61	60	60	60	59	59	59	58	58	58	57
52	51	50	48	47	46	45	44	43	42	41	40	39	38	37	36	35	35	34	33	32	32	31	30	30	29	29	28	28	27
61	60	59	58	57	56	55	54	53	52	51	50	49	48	48	47	46	45	45	44	43	43	42	41	41	40	40	39	38	38
71	70	69	68	68	67	66	66	65	64	64	63	62	62	61	60	60	59	59	58	57	57	56	56	55	55	54	54	53	53
83	83	83	82	82	82	82	81	81	81	80	80	80	80	79	79	79	79	78	78	78	78	78	77	77	77	77	76	76	76

Ceiling or Floor Cavity Ratio

The space between fixture and ceiling is the *ceiling cavity*. The space between the work plane and the floor is the *floor cavity*. The space between the fixture and the work plane is the *room cavity*. To determine the room cavity ratio use Figure 5-1 to define the cavity depth and then find the corresponding ratio in Table 5-5.

To determine the effective ceiling cavity reflectance, proceed in the same manner to define the ceiling cavity ratio, then refer to Table 5-6 to find the corresponding effective ceiling cavity reflectance.

SIM 5-3

For Process Plant No. 1, determine the coefficient of utilization for a room which measures 24' X 100'. The ceiling is 20' high and the fixture is mounted 4' from the ceiling. Use the data in Table 5-3.

Answer

Step (a)

Since no wall or ceiling reflectance data was given, assume a ceiling of .70 and wall of .5.

Step (b)

Assume 3' working height.
hrc = 20-4-3 = 13 (from Figure 5-1)
From Table 5-5, RCR = 3.4

Step (c)

From Figure 5-1, hcc = 4
From Table 5-5, CCR = 1
From Table 5-6, pCC = 58
From Table 5-3, Coefficient of Utilization =
                .64 (interpolated)

THE LIGHT LOSS FACTOR

The light loss factor (LLF) takes into account that the lamp lumen depreciates with time (L_1), that the lumen output depreciates due to dirt build-up (L_2), and that lamps burn out (L_3). Formula 5-2 illustrates the relationship of these factors.

(Formula 5-2) $$LLF = L_1 \times L_2 \times L_3$$

To reduce the number of lamps required which in turn reduces energy consumption, it is necessary to increase the overall light loss factor. This is accomplished in several ways. One is to choose the luminaire which minimizes dust build-up. The second is to improve the maintenance program to replace lamps prior to burn-out. Thus if it is known that a group relamping program will be used at a given percentage of rated life, the appropriate lumen depreciation factor can be found from Table 5-2. It may be decided to use a shorter relamping period in order to increase (L_1) even further. If a group relamping program is used (L_3) is assumed to be unity.

Figure 5-2 illustrates the effect of dirt build-up on (L_2) for a dustproof luminaire. Every luminaire has a tendency for dirt build-up. Manufacturer's data should be consulted when estimating L_2 for the luminaire in question.

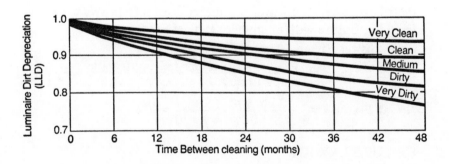

Figure 5-2. Effect of Dirt Build-Up on Dustproof Luminaires for Various Atmospheric Conditions

SIM 5-4

Estimate the light loss factor for the following cases.

Case 1 — group relamping at 8000 hours.
Case 2 — group relamping at 12,000 hours.

In both cases the fixture lens is wiped when relamped. Assume a lamp burn-out factor of unity. Use the data for the Deluxe white (175W) mercury vapor lamp illustrated in Table 5-2 and the dirt accumulation factor of Figure 5-2. Plant operates 8,000 hours per year with a dirty environment.

Answer

Case 1 — L_1 = .86 (Table 5-2)
L_2 = .92 (Figure 5-2)
LLF = .86 X .92 X 1 = .79

Case 2 — L_1 = .81
L_2 = .9
LLF = .81 X .9 X 1 = .72

Thus the number of lamps required would be approximately 9% less if a group relamping program were initiated at 8,000 hours of lamp life.

SIM 5-5

For SIM 5-3 determine the required number of fixtures.

Desired footcandle level is 50. The light loss factor is estimated to be .7. The lamps are mercury vapor, 175W, with one lamp per fixture with a lumen output of 8500.

Answer (from Formula 5-1)

$$\text{No. of Fixtures} = \frac{\text{Area X Desired Footcandle}}{\text{Lamps/Fixt. X Lumens/Lamp X C.U. X LLF}}$$

$$\text{No. of Fixtures} = \frac{24 \text{ X } 100 \text{ X } 50}{1 \text{ X } 8500 \text{ X } .64 \text{ X } .70} = 32$$

FIXTURE LAYOUT

The fixture layout is dependent on the area. The initial layout should have equal spacing between lamps, rows and columns. The end fixture should be located at one-half the distance between fixtures. The maximum distance between fixtures usually should not exceed the mounting height unless the manufacturer specifies otherwise. Figure 5-3 illustrates a typical layout. If the fixture is fluorescent, it may be more practical to run the fixtures together. Since the fixtures are 4 feet or 8 feet long, a continuous wireway will be formed.

**Figure 5-3.
Typical Fixture
Layout**

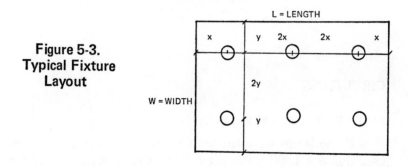

SIM 5-6

For SIM 5-5 design a lighting layout.

Answer

From SIM 5-5, thirty-two 175 watt mercury vapor lamps are required.

		Rows	Columns	X Spacing	Y Spacing
Typical					
Combinations	(a)	4	8	12.5	6
	(b)	3	11	9	8
	(c)	2	16	6	12

(a)	$8X = 100$	(b)	$11X = 100$	(c)	$16X = 100$
	$X = 12.5$		$X = 9$		$X = 6.2$
	$4Y = 24$		$3Y = 24$		$2Y = 24$
	$Y = 6$		$Y = 8$		$Y = 12$

Alternate (b) is recommended even though it requires one more fixture. It results in a good layout, illustrated following.

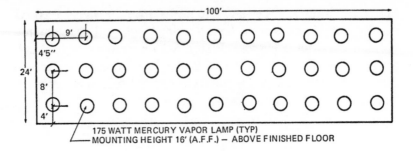

175 WATT MERCURY VAPOR LAMP (TYP)
MOUNTING HEIGHT 16' (A.F.F.) — ABOVE FINISHED FLOOR

CIRCUITING

NUMBER OF LAMPS PER CIRCUIT

If the lamps were fed from a 110-volt source, a commonly used guide is 1600 watts per lighting circuit breaker (based on 20 amp breaker, #12 wire.) This load includes fixture voltage and ballast loss. In SIM 5-6, assuming a ballast loss of 25 watts per fixture, eight lamps could be fed from each circuit breaker.

A single-phase circuit panel is illustrated in Figure 5-4. (Note: In practice ballast loss should be based on man-ufacturer's specifications.)

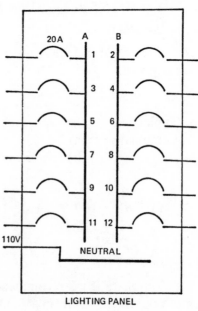

**Figure 5-4.
Single-Phase
Circuit Panel**

LIGHTING PANEL

SIM 5-7

Next to each lamp place the circuit number from which each lamp is fed; i.e., A-1, A-2, etc.

| ○ | ○ | ○ | ○ | ○ | ○ | ○ | ○ | ○ | ○ | ○ |

Answer

A-1	A-1	A-1	A-1	A-1	A-1	A-1	A-1	A-4	A-4	A-4
○	○	○	○	○	○	○	○	○	○	○
A-2	A-2	A-2	A-2	A-2	A-2	A-2	A-2	A-4	A-4	A-4
○	○	○	○	○	○	○	○	○	○	○
A-3	A-3	A-3	A-3	A-3	A-3	A-3	A-3	A-4	A-4	A-5
○	○	○	○	○	○	○	○	○	○	○

SIM 5-8

Designate a hot line from the circuit breaker with a small stroke and use a long stroke as a neutral; i.e., ⎯⫽⫽⎯ 4 wires, 2 hot and 2 neutrals. The lamps are connected with conduit as shown below. Designate the hot and neutrals in each branch.

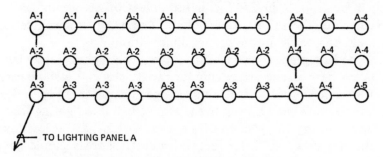

Hint—start wiring from the last fixture in the circuit.

Answer

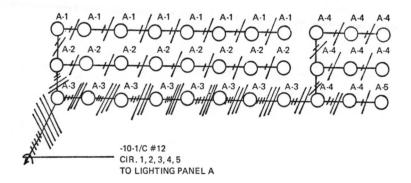

POINTS ON LIGHTING DRAWINGS

• Choose a lighting drawing scale based on the area to be lighted and the detail required. Typical drawing scale: 1/8" equals one foot.

• Identify all symbols for lighting fixtures.

• Include circuit numbers on all lights.

• Include a note on fixture mounting height.

• Show "homerun" to lighting panels. "Homerun" indicates the number of wires and conduit size from the last outlet box.

• Use notes to simplify drawing. For example: All wires shall be 2 #12 in 3/4" conduit unless otherwise indicated. Remember the information put on a drawing or specification should be clear to insure proper illumination.

• Receptacles are usually shown on lighting drawings. Always have a separate circuit for receptacles and assume each outlet consumes 200 watts unless specific uses are known. Receptacle wires may be run in the same conduit as lighting wires. Manufacturing and general office areas should have at least one outlet for every 20 linear feet. Private office areas should have at least one outlet for every 10 linear feet.

JOB SIMULATION – SUMMARY PROBLEM

JOB 4

(a)   The Ajax Plant, Job 1, Chapter 2, contains a basement and an operating floor; each an area of 20' X 400'.

For this plant compute the number of lamps for the basement area, the space between fixtures, and the number of circuits. Use two 40-watt fluorescent lamps per fixture, 2900 lumens per lamp, light loss factor = .7, 110-volt lighting system, 20-watt ballast loss per fixture, and a fixture length of 2' X 4'. Use luminaire data of Table 5-7, ceiling height 20' and a desired footcandle level of 40.

When calculating RCR, assume that the total area is divided into four equal rooms per floor, approximately 100' long.

Table 5-7. Coefficient of Utilization
20% Effective Floor Cavity Reflectance

Effective Ceiling Cavity Reflectance	80%			50%		
Wall Reflectance	50	30	10	50	30	10
RCR						
10	.33	.26	.22	.31	.26	.22
9	.43	.35	.27	.40	.35	.29
8	.58	.42	.35	.48	.42	.36
7	.58	.50	.42	.55	.48	.42
6	.64	.57	.49	.61	.54	.47
5	.72	.65	.59	.65	.60	.56
4	.77	.71	.64	.71	.65	.60
3	.82	.76	.70	.74	.69	.63
2	.87	.82	.77	.78	.74	.70
1	.91	.87	.83	.81	.78	.75
Spacing not to exceed 1 X Mounting Height						

Analysis

(a)   Total Area – 20' X 400'

Since the problem stated four equal rooms approximately 100′ long, use the area 20′ X 100′ to calculate the coefficient of utilization.

Assume hfc = 3
hcc = 3
hrc = 14 based on 20′ ceiling.

The room cavity ratio is approximately 4.2, based on 90′ length. The ceiling cavity ratio is .9.

Assume 70% ceiling reflectance
50% wall reflectance

Thus the effective ceiling reflectance pcc = 59. Fifty per cent is the lowest value given in the vendor's data. Since the C.U. does not seem to vary too much with pcc, use pcc = 50 to determine the C.U.

Thus C.U. = .71

$$\text{No. of Fixtures} = \frac{20 \times 100 \times 40}{2 \times 2900 \times .71 \times .70} = 28$$

Layout per Room:

(1) 2 rows
14 columns
14y = 100
y = 7.1
2x = 20
x = 10

(2) 3 rows
9 columns
9y = 100
y = 11
3x = 20
x = 6.6

2 rows and 14 columns seem to offer the best spacing.

Total number of fixtures required:

28 fixtures/room X 4 rooms/floor X 2 floors = 224 fixtures. Basement area requires 112 fixtures.

(b) The area of the electric room is 18′-6″ X 20′.

with hfc = 3
hcc = 3
hrc = 14

The room cavity ratio is 7 and the effective ceiling cavity ratio pcc = 53. Thus C.U. = .55.

$$\text{No. of Fixtures} = \frac{20' \times 18\frac{1}{2}' \times 40}{2 \times 2900 \times .55 \times .70} = 7$$

Layout Spacing

3x = 20
 x = 6.6
3y = 18½'
 y = 6'2"

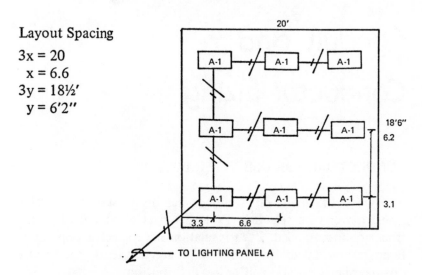

Note: With emphasis on energy conservation, a lighting layout using 6 fixtures may be preferable.

SUMMARY

Energy conservation is influencing lighting design. Increased emphasis is being placed on minimizing lighting fixtures by using luminaires which have a high lumen output and coefficient of utilization. Today's lighting systems incorporate switching and automatic control devices to make it easy to turn off lights when they are not required. Lighting systems need to be analyzed on a first and operating cost basis to insure that the increasing energy costs are taken into account.

In addition to the general lighting methods outlined more attention should be placed on providing light where needed. This will mean using the "point by point" lighting method and using unsymmetrical lighting layouts in order to minimize energy consumption.

6

Conduit and
Conductor Sizing

CONDUCTORS FOR USE IN RACEWAYS

Once the plant environment has been established, cable insulation can be selected. Table 6-1 lists a number of insulations commonly used. Each insulation has a specified operating temperature. Types THW and RHW are frequently used in industrial plants for 600 volt or less distribution.

Allowable ampacities of copper conductors in conduit at 30°C are shown in Table 6-2. Sometimes the designations 0, 00, 0000 are referred to as 1/0, 2/0, 3/0, 4/0, respectively.

Tables 6-2 through 6-7 are reproduced by permission from the National Electrical Code 1978, copyright 1977 National Fire Protection Association. Copies of the complete code are available from the Association, 470 Lexington Avenue, Boston, Massachusetts 02110.

SIM 6-1

For one 3/C 350 MCM THW copper cable, run in conduit at 30°C, determine the cable capacity.

Answer

From Table 6-2 type THW is rated for 310 amperes.

Table 6-1. Conductors for Use in Raceways
(A Partial Listing)

TYPE	INSULATION
FEP	Fluorinated Ethylene Propylene
MI	Mineral Insulation (metal sheathed)
RH	Heat-Resistant Rubber
RHH	Heat-Resistant Rubber
RHW	Moisture and Heat-Resistant Rubber
RUH	Heat-Resistant Latex Rubber
RUW	Moisture-Resistant Latex Rubber
T	Thermoplastic
TW	Moisture-Resistant Thermoplastic
THW	Moisture and Heat-Resistant Thermoplastic
THWN	Moisture and Heat-Resistant Thermoplastic
THHN	Moisture and Heat-Resistant Thermoplastic
XHHW	Moisture and Heat Resistant Cross-Linked Synthetic Polymer
V	Varnished Cambric
AVA	Asbestos and Varnished Cambric
AVL	Asbestos and Varnished Cambric
AVB	Asbestos and Varnished Cambric

SIM 6-2

What type of insulation is type RHW?

Answer

From Table 6-1 RHW is composed of moisture and heat resistant rubber.

SIM 6-3

Determine the ampacity for one 3/C 4/0 cross-linked polyethylene copper cable run in conduit at 86°F in a dry location.

Answer

In Table 6-1 cross-linked polyethylene copper cable is Type XHHW. When it is used in a dry location it is good for 90°C. From Table 6-2, 0000 is good for 235 amperes.

Table 6-2.

NEC Table 310-16. Allowable Ampacities of Insulated Conductors Rated 0-2000 Volts, 60° to 90°C

Not More Than Three Conductors in Raceway or Cable or Earth
(Directly Buried), Based on Ambient Temperature of 30°C (86°F)

Size	Temperature Rating of Conductor. See Table 310-13								Size
	60°C (140°F)	75°C (167°F)	85°C (185°F)	90°C (194°F)	60°C (140°F)	75°C (167°F)	85°C (185°F)	90°C (194°F)	
AWG MCM	TYPES RUW, T, TW, UF	TYPES FEPW, RH, RHW, RUH, THW, THWN, XHHW, USE, ZW	TYPES V, MI	TYPES TA, TBS, SA, AVB, SIS, †FEP, †FEPB, †RHH, †THHN, †XHHW*	TYPES RUW, T, TW, UF	TYPES RH, RHW, RUH, THW, THWN, XHHW, USE	TYPES V, MI	TYPES TA, TBS, SA, AVB, SIS, †RHH, †THHN, †XHHW*	AWG MCM
	COPPER				ALUMINUM OR COPPER-CLAD ALUMINUM				
18	....	....	....	21	....	....	....	....	....
16	....	....	22	22	....	....	....	....	....
14	15	15	25	25	....	....	....	....	....
12	20	20	30	30	15	15	25	25	12
10	30	30	40	40	25	25	30	30	10
8	40	45	50	50	30	40	40	40	8
6	55	65	70	70	40	50	55	55	6
4	70	85	90	90	55	65	70	70	4
3	80	100	105	105	65	75	80	80	3
2	95	115	120	120	75	90	95	95	2
1	110	130	140	140	85	100	110	110	1
0	125	150	155	155	100	120	125	125	0
00	145	175	185	185	115	135	145	145	00
000	165	200	210	210	130	155	165	165	000
0000	195	230	235	235	155	180	185	185	0000
250	215	255	270	270	170	205	215	215	250
300	240	285	300	300	190	230	240	240	300
350	260	310	325	325	210	250	260	260	350
400	280	335	360	360	225	270	290	290	400
500	320	380	405	405	260	310	330	330	500
600	355	420	455	455	285	340	370	370	600
700	385	460	490	490	310	375	395	395	700
750	400	475	500	500	320	385	405	405	750
800	410	490	515	515	330	395	415	415	800
900	435	520	555	555	355	425	455	455	900
1000	455	545	585	585	375	445	480	480	1000
1250	495	590	645	645	405	485	530	530	1250
1500	520	625	700	700	435	520	580	580	1500
1750	545	650	735	735	455	545	615	615	1750
2000	560	665	775	775	470	560	650	650	2000
CORRECTION FACTORS									

Ambient Temp. °C	For ambient temperatures over 30°C, multiply the ampacities shown above by the appropriate correction factor to determine the maximum allowable load current.								Ambient Temp. °F
31-40	.82	.88	.90	.91	.82	.88	.90	.91	86-104
41-50	.58	.75	.80	.82	.58	.75	.80	.82	105-122
51-60	....	.58	.67	.71	....	.58	.67	.71	123-141
61-70	....	.35	.52	.58	....	.35	.52	.58	142-158
71-80	....	....	.30	.41	....	....	.30	.41	159-176

† The load current rating and the overcurrent protection for these conductors shall not exceed 15 amperes for 14 AWG, 20 amperes for 12 AWG, and 30 amperes for 10 AWG copper; or 15 amperes for 12 AWG and 25 amperes for 10 AWG aluminum and copper-clad aluminum.

* For dry locations only. See 75°C column for wet locations.

Table 6-2. (concluded)   155

NEC
Table 310-13 (Continued)

Trade Name	Type Letter	Max. Operating Temp.	Application Provisions	Insulation	AWG or MCM	Thickness of Insulation	Mils	Outer Covering
Asbestos	A	200°C 392°F	Dry locations only. Only for leads within apparatus or within raceways connected to apparatus. Limited to 300 volts.	Asbestos	14 12-8		30 40	Without asbestos braid
Asbestos	AA	200°C 392°F	Dry locations only. Only for leads within apparatus or within raceways connected to apparatus or as open wiring. Limited to 300 volts.	Asbestos	14 12-8 6-2 1-4/0		30 30 40 60	With asbestos braid or glass
Asbestos	AI	125°C 257°F	Dry locations only. Only for leads within apparatus or within raceways connected to apparatus. Limited to 300 volts.	Impregnated Asbestos	14 12-8		30 40	Without asbestos braid
Asbestos	AIA	125°C 257°F	Dry locations only. Only for leads within apparatus or within raceways connected to apparatus or as open wiring.	Impregnated Asbestos	14 12-8 6-2 1-4/0 213-500 501-1000		Sol. Str. 30 30 40 40 60 60 75 90 105	With asbestos braid or glass
Paper		85°C 185°F	For underground service conductors, or by special permission.	Paper				Lead sheath

For insulated aluminum and copper-clad aluminum conductors, the minimum size shall be No. 12. See Tables 310-16 through 310-19.

CONDUCTOR DERATINGS

DERATING DUE TO THE NUMBER OF CONDUCTORS

If three conductors or less are run in a conduit, no derating need be applied. Based on the number of conductors above three, additional deratings must be used. For example: 4 to 6 conductors derate the cable 80%; 7 to 24, 70%; 25 to 42, 60%; and 43 and above, 50%. Because of these deratings and ease in pulling, many times six conductors servicing a load are run in two conduits.

DERATING DUE TO AMBIENT

If the ambient differs from 30°C, another derating should be applied. These correction factors are included in Table 6-2.

SIM 6-4

What is the allowable ampacity of two 3/C, 500 MCM, THW, copper conductors run in the same conduit at 50°C?

Answer

From Table 6-2, the allowable ampacity of 500 MCM cable is 380 amperes.

Derating factor for six conductors is .8.

Allowable ampacity = 2 X (380 X .8 X .75) =
228 X 2 amp = 456 amps.

OBSERVATION

It should be noted from Table 6-2 that some conductor sizes are not practical. Doubling the area of a conductor, i.e., 500 MCM cable to a 1000 MCM cable, does not double the ampacity rating. In fact doubling the area only results in 50% more capacity. Using conductors larger than 500 MCM is also not recommended due to difficulties in installation.

CABLE SIZING

SIZING A FEEDER TO A MOTOR FOR CONTINUOUS SERVICE

The minimum size cable for power conductors is #12. The cable capacity for a motor should be equal to 1.25 X the full load amperes of the motor. Typical full-load currents are illustrated in Chapter 4, Table 4-2.

SIM 6-5

What cable size should be used for a 50 HP induction motor? Assume the cable is Type THW and run in conduit. Ambient temperature 30°C.

Answer

From Table 4-2, Chapter 4, a 50 HP motor has a F.L.A. (full load amperes) of 65 amps. Cable must be equal to or above 1.25 X 65 = 81 amp. Use #3/C #4 AWG.

SIZING FEEDER TO SEVERAL MOTORS

The size of the feeder which has more than one motor on it is based on 1.25 times the full-load current of the largest motor plus the full-load current of the others.

SIM 6-6

A feeder supplies a 25 HP motor and two 30 HP motors. Determine the size of the feeder if it is rated for 60°C. Ambient 30°C and cable run in conduit.

Answer

F.L.A. – 30 HP motor – 40 amp.
F.L.A. – 25 HP motor – 34 amp.
Cable must meet: 1.25 X 40 + 34 + 40 = 124
The nearest cable size from Table 6-2 is #1 THW.

SIZING A CONDUIT

The size of a conduit depends on the allowable percent fill of the conduit area. Table 6-3 illustrates allowable fills. Table 6-4 is based on Table 6-3 and enables the engineer to readily select the number of conductors which can be installed in a conduit.

Table 6-3.

NEC Table 1.   Percent of Cross Section of Conduit and Tubing for Conductors

(See Table 2 for Fixture Wires)

Number of Conductors	1	2	3	4	Over 4
All conductor types except lead-covered (new or rewiring)	53	31	40	40	40
Lead-covered conductors	55	30	40	38	35

Reproduced with permission from the National Electrical Code, 1978 edition, copyright 1977, National Fire Protection Association, 470 Lexington Avenue, Boston, MA 02210.

Table 6-4 stipulates how many of the same size cables can be run in a conduit based on the allowable fills mentioned in Table 6-3.

SIM 6-7

What is the allowable percent fill for three conductors Type THW run in conduit.

Answer

From Table 6-3, 40%.

SIM 6-8

A control cable consisting of thirty-seven 1/C #14 wires Type THW is run from M.C.C. No. 1 to Process Panel No. 2. What size conduit is required?

Answer

From Table 6-3 the maximum conduit fill is 40%. From Table 6-4 the answer can be read directly. Since 40 wires can be run in a 1½" conduit, the correct size is 1½".

Table 6-4

Reproduced with permission from the National Electrical Code, 1978 edition, copyright 1977, National Fire Protection Association, 470 Lexington Avenue, Boston, MA 02210.

NEC Table 3A. Maximum Number of Conductors in Trade Sizes of Conduit or Tubing
(Based on Table 1, Chapter 9)

Type Letters	Conductor Size AWG, MCM	½	¾	1	1¼	1½	2	2½	3	3½	4	4½	5	6
TW, T, RUH, RUW, XHHW (14 thru 8)	14	9	15	25	44	60	99	142	171					
	12	7	12	19	35	47	78	111	131	176				
	10	5	9	15	26	36	60	85						
	8	2	4	7	12	17	28	40	62	84	108			
RHW and RHH (without outer covering), THW	14	6	10	16	29	40	65	93	143	192				
	12	4	8	13	24	32	53	76	117	157				
	10	4	6	11	19	26	43	61	95	127	163			
	8	1	3	5	10	13	22	32	49	66	85	106	133	
TW, T, THW, RUH (6 thru 2), RUW (6 thru 2)	6	1	2	4	7	10	16	23	36	48	62	78	97	141
	4	1	1	3	5	7	12	17	27	36	47	58	73	106
	3	1	1	2	4	6	10	15	23	31	40	50	63	91
	2		1	2	4	5	9	13	20	27	34	43	54	78
	1			1	3	4	6	9	14	19	25	31	39	57
FEPB (6 thru 2), RHW and RHH (without outer covering)	0		1	1	2	3	5	8	12	16	21	27	33	49
	00		1	1	1	3	5	7	10	14	18	23	29	41
	000		1	1	1	2	4	6	9	12	15	19	24	35
	0000			1	1	1	3	5	7	10	13	16	20	29
(RHW and RHH without outer covering)	250			1	1	1	2	4	6	8	10	13	16	23
	300			1	1	1	2	3	5	7	9	11	14	20
	350			1	1	1	1	3	4	6	8	10	12	18
	400				1	1	1	2	4	5	7	9	11	16
	500				1	1	1	1	3	4	6	7	9	14
	600					1	1	1	3	4	5	6	7	11
	700					1	1	1	2	3	4	5	7	10
	750					1	1	1	2	3	4	5	6	9

/more/

Table 6-4. (concluded)

NEC Table 3B
Maximum Number of Conductors in Trade Sizes of Conduit or Tubing
(Based on Table 1, Chapter 9)

Type Letters	Conductor Size AWG, MCM	½	¾	1	1¼	1½	2	2½	3	3½	4	4½	5	6
THWN,	14	13	24	39	69	94	154	164	160	106	136			
	12	10	18	29	51	70	114	104	79					
	10	6	11	18	32	44	73	51						
	8	3	5	9	16	22	36							
THHN, FEP (14 thru 2), FEPB (14 thru 8), PFA (14 thru 4/0), PFAH (14 thru 4/0), Z (14 thru 4/0),	6	1	4	6	11	15	26	37	57	76	98	125	154	137
	4	1	2	4	7	9	16	22	35	47	60	75	94	116
	3	1	1	3	6	8	13	19	29	39	51	64	80	97
	2	1	1	3	5	7	11	16	25	33	43	54	67	72
	1		1	1	3	5	8	12	18	25	32	40	50	
XHHW (4 thru 500 MCM)	0	1	1	1	3	4	7	10	15	21	27	33	42	61
	00	1	1	1	2	3	6	8	13	17	22	28	35	51
	000	1	1	1	1	3	5	7	11	14	18	23	29	42
	0000				1	2	4	6	9	12	15	19	24	35
	250			1	1	1	3	4	7	10	12	16	20	28
	300			1	1	1	3	4	6	8	11	13	17	24
	350			1	1	1	2	3	5	7	9	12	15	21
	400				1	1	1	3	5	6	8	10	13	19
	500				1	1	1	2	4	5	7	9	11	16
	600				1	1	1	1	3	4	5	7	9	13
	700					1	1	1	3	4	5	6	8	11
	750					1	1	1	2	3	4	6	7	11
XHHW	6	1	3	5	9	13	21	30	47	63	81	102	128	185
	600				1	1	1	1	3	4	5	7	9	13
	700					1	1	1	3	4	5	6	7	11
	750					1	1	1	2	3	4	6	7	10

Table 6-5.

Reproduced with permission from the National Electrical Code, 1978 edition, copyright 1977, National Fire Protection Association, 470 Lexington Avenue, Boston, MA 02210.

NEC Table 5. Dimensions of Rubber-Covered and Thermoplastic-Covered Conductors

Size AWG MCM	Types RFH-2, RH, RHH,*** RHW,*** SF-2 Approx. Diam. Inches	Approx. Area Sq. In.	Types TF, T, THW,† TW, RUH,** RUW** Approx. Diam. Inches	Approx. Area Sq. In.	Types TFN, THHN, THWN Approx. Diam. Inches	Approx. Area Sq. In.	Types **** FEP, FEPB, FEPW, TFE, PF, PFA, PFAH, PGF, PTF, Z, ZF, ZFF Approx. Diam. Inches	Approx. Area Sq. Inches	Type XHHW, ZW†† Approx. Diam. inches	Approx. Area Sq. In.	Types KF-1, KF-2, KFF-1, KFF-2 Approx. Diam. Sq. In.	Approx. Area Sq. In.
Col. 1	Col. 2	Col. 3	Col. 4	Col. 5	Col. 6	Col. 7	Col. 8	Col. 9	Col. 10	Col. 11	Col. 12	Col. 13
18	.146	.0167	.106	.0088	.089	.0064	.081	.0052	...	....	.065	.0033
16	.158	.0196	.118	.0109	.100	.0079	.092	.0066	...	....	.070	.0038
14	30 mils .171	.0230	.131	.0135	.105	.0087	.105	.0087			.083	.0054
14	45 mils .204*	.0327*	.162†	.0206†	...	...	.105	.0087	.129	.0131		
12	30 mils .188	.0278	.148	.0172	.122	.0117	.121	.0115			.102	.0082
12	45 mils .221*	.0384*	.179†	.0251†	...	...	.121	.0115	.146	.0167		
10	.242	.0460	.168	.0224	.153	.0184	.142	.0159			.124	.0121
10			.199†	.0311†	...	...	.142	.0159	.166	.0216		
8	.328	.0854	.245	.0471	.218	.0373	.206	.0333				
8		....	.276†	.0598†	...	...	.186	.0272	.241	.0456		
6	.397	.1238	.323	.0819	.257	.0519	.244	.0467	.282	.0625		
							.302	.0716				
4	.452	.1605	.372	.1087	.328	.0845	.292	.0669	.328	.0845		
							.350	.0962				
3	.481	.1817	.401	.1263	.356	.0995	.320	.0803	.356	.0995		
							.378	.1122				
2	.513	.2067	.433	.1473	.388	.1182	.352	.0973	.388	.1182		
							.410	.1316				
1	.588	.2715	.508	.2027	.450	.1590	.420	.1385	.450	.1590		
0	.629	.3107	.549	.2367	.491	.1893	.462	.1676	.491	.1893		
00	.675	.3578	.595	.2781	.537	.2265	.498	.1974	.537	.2265		
000	.727	.4151	.647†	.3288	.588	.2715	.560	.2463	.588	.2715		
0000	.785	.4840	.705	.3904	.646	.3278	.618	.2999	.646	.3278		

/more/

NEC Table 5 (Continued)

Table 6-5. (concluded)

Size AWG MCM	Types RFH-2, RH, RHH,*** RHW,*** SF-2		Types TF, T, THW,† TW, RUH,** RUW**		Types TFN, THHN, THWN		Types **** FEP, FEPB, FEPW, TFE, PF, PFA, PFAH, PGF, PTF, Z, ZF, ZFF		Type XHHW, ZW††	
	Approx. Diam. Inches	Approx. Area Sq. In.	Approx. Diam. Inches	Approx. Area Sq. In.	Approx. Diam. Inches	Approx. Area Sq. In.	Approx. Diam. Inches	Approx. Area Sq. Inches	Approx. Diam. Inches	Approx. Area Sq. In.
Col. 1	Col. 2	Col. 3	Col. 4	Col. 5	Col. 6	Col. 7	Col. 8	Col. 9	Col. 10	Col. 11
250	.868	.5917	.788	.4877	.716	.4026	...	...	.716	.4026
300	.933	.6837	.843	.5581	.771	.4669	...	...	.771	.4669
350	.985	.7620	.895	.6291	.822	.5307	...	...	.822	.5307
400	1.032	.8365	.942	.6969	.869	.5931	...	...	.869	.5931
500	1.119	.9834	1.029	.8316	.955	.7163	...	...	.955	.7163
600	1.233	1.1940	1.143	1.0261	1.058	.8792	...	...	1.073	.9043
700	1.304	1.3355	1.214	1.1575	1.129	1.0011	...	...	1.145	1.0297
750	1.339	1.4082	1.249	1.2252	1.163	1.0623	...	...	1.180	1.0936
800	1.372	1.4784	1.282	1.2908	1.196	1.1234	...	...	1.210	1.1499
900	1.435	1.6173	1.345	1.4208	1.259	1.2449	...	...	1.270	1.2668
1000	1.494	1.7531	1.404	1.5482	1.317	1.3623	...	...	1.330	1.3893
1250	1.676	2.2062	1.577	1.9532	...	...	...	...	1.500	1.7672
1500	1.801	2.5475	1.702	2.2748	...	...	...	...	1.620	2.0612
1750	1.916	2.8895	1.817	2.5930	...	...	...	...	1.740	2.3779
2000	2.021	3.2079	1.922	2.9013	...	...	...	...	1.840	2.6590

* The dimensions of Types RHH and RHW.
** No. 14 to No. 2.
† Dimensions of THW in sizes 14 to 8. No. 6 THW and larger is the same dimension as T.
*** Dimensions of RHH and RHW without outer covering are the same as THW.
No. 18 to No. 10, solid; No. 8 and larger, stranded.
**** In Columns 8 and 9 the values shown for sizes No. 1 thru 0000 are for TFE and Z only. The right-hand values in Columns 8 and 9 are for FEPB, Z, ZF, and ZFF only.
†† No. 14 to No. 2.

POWER AND CONTROL RUN IN THE SAME CONDUIT

Frequently the electrical engineer is faced with the problem of sizing a conduit for both power and control wires. An example might be a local STOP-START pushbutton at the motor. The control wires would be run in the same conduit as the power leads if the horsepower is 60 HP or less. Above this size, it becomes impractical to pull the smaller control wires with the larger power conductors. For different size cables, first use Table 6-5 to determine the area of each cable. Then look at Table 6-6 under the appropriate percent fill column and choose a conduit size whose area is equal to or greater than the total conductor area.

SIM 6-9

Determine the conduit size for the 50 HP motor, SIM 6-5. Assume a local stop-start pushbutton requiring three 1/C #14 control cables.

Answer

From SIM 6-5 a 3/C #4 cable is required for a 50 HP motor. Table 6-6 shows cable areas for single conductors. Multiply the area by 3 to get the total area. This approximation gives a conservative cable sizing.

$$
\begin{array}{lll}
\text{Power} & 3 \times .1087 = & .3261 \\
\text{Control} & 3 \times .0206 = & \underline{.0618} \\
& \text{Total} & .3879
\end{array}
$$

From Table 6-6 based on 40% fill, conduit size is 1¼".

SUMMARY OF DATA

At the beginning of each project a table should be established summarizing all conduit, cable, fuse and switch or breaker sizes for each motor horsepower. At a glance each motor and its associated auxiliaries can be determined. On a large project this type of table saves a considerable amount of time.

Table 6-6.

Reproduced with permission from the National Electrical Code, 1978 edition, copyright 1977, National Fire Protection Association, 470 Lexington Avenue, Boston, MA 02210.

NEC Table 4.   Dimensions and Percent Area of Conduit and of Tubing

Areas of Conduit or Tubing for the Combinations of Wires Permitted in Table 1, Chapter 9.

| Trade Size | Internal Diameter Inches | Area — Square Inches | | | | | | | | |
| | | Total 100% | Not Lead Covered | | | Lead Covered | | | | |
			2 Cond. 31%	Over 2 Cond. 40%	1 Cond. 53%	1 Cond. 55%	2 Cond. 30%	3 Cond. 40%	4 Cond. 38%	Over 4 Cond. 35%
½	.622	.30	.09	.12	.16	.17	.09	.12	.11	.11
¾	.824	.53	.16	.21	.28	.29	.16	.21	.20	.19
1	1.049	.86	.27	.34	.46	.47	.26	.34	.33	.30
1¼	1.380	1.50	.47	.60	.80	.83	.45	.60	.57	.53
1½	1.610	2.04	.63	.82	1.08	1.12	.61	.82	.78	.71
2	2.067	3.36	1.04	1.34	1.78	1.85	1.01	1.34	1.28	1.18
2½	2.469	4.79	1.48	1.92	2.54	2.63	1.44	1.92	1.82	1.68
3	3.068	7.38	2.29	2.95	3.91	4.06	2.21	2.95	2.80	2.58
3½	3.548	9.90	3.07	3.96	5.25	5.44	2.97	3.96	3.76	3.47
4	4.026	12.72	3.94	5.09	6.74	7.00	3.82	5.09	4.83	4.45
4½	4.506	15.94	4.94	6.38	8.45	8.77	4.78	6.38	6.06	5.56
5	5.047	20.00	6.20	8.00	10.60	11.00	6.00	8.00	7.60	7.00
6	6.065	28.89	8.96	11.56	15.31	15.89	8.67	11.56	10.98	10.11

POWER LAYOUTS

Power layouts are drawn to scale (usually ¼″ = 1′). Conduits are grouped together where possible to form conduit banks. Usually conduits run vertically and horizontally. If conduit and cable sizes cannot be shown on drawing, a separate conduit and cable schedule is required. Conduit layouts should be coordinated with other groups (Piping, HVAC) to avoid interferences.

SIM 6-10

For the plan below, draw a conduit layout.

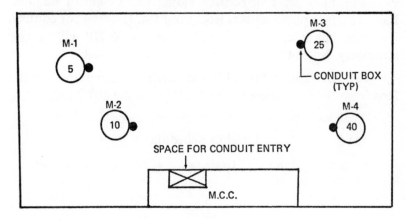

Answer

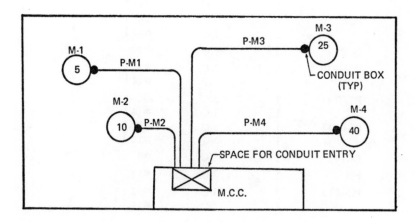

JOB SIMULATION – SUMMARY PROBLEM

JOB 5

(a)   With the motor data of Job 1, Chapter 2, establish a conduit and cable schedule. Ambient 40°C and wire Type THW. Assume each stop-start motor requires three 1/C #14 for a local pushbutton station. No local control is provided for the two-speed motor, the reversing motors, and AG-1 and FP4. Assume 3/C only to hoist for power. Control by others.

When power and control are run in the same conduit, designate the conduit by PC–motor number. For power alone, use P–motor number. For control alone, use C–motor number. Exclude single-phase motors from your list. Use #14 for control.

Remember control for motors above 60 HP will be run separately. Note: Cable size 3 is not frequently used.

(b)   Determine the cable and conduit size required for the feed from the substation to the M.C.C. Assume 40°C ambient.

Analysis

(a)   **Conduit Schedule**

Conduit No.	No.	Conductors	Size	Conduit Size
P-AG1	1	3/C	#2	1¼"
P-CF3	1	3/C	#210	1½"
P-FP4	1	3/C	#6	1"
PC-TP-5	1	3/C	#12 ⎱	¾"
	1	3/C	#14 ⎰	
PC-CTP6	1	3/C	#6 ⎱	1"
	1	3/C	#14 ⎰	
P-CT9	2	3/C	#8	1"
PC-HF10	1	3/C	#4 ⎱	1¼"
	1	3/C	#14 ⎰	
PC-HF11	1	3/C	#8 ⎱	1"
	1	3/C	#14 ⎰	
PC-BC13	1	3/C	#2 ⎱	1¼"
	1	3/C	#14 ⎰	
P-C16	1	3/C	#8	¾"
P-H17	1	3/C	#12	¾"
PC-SC19	1	3/C	#12 ⎱	¾"
	1	3/C	#14 ⎰	

(b)   The cable to the M.C.C. should be at least 1.25% of the F.L.A. of the largest motor plus the full load amps of the others. In this case it is approximately 525 amps. The lighting and fractional HP load of approximately 15 amps is then added, bringing the total to 540 amps.

Two 350 MCM run in separate conduits can feed a total load of 2 X 310 X .88 = 545.

For this application, two 2½ conduits are required.

SUMMARY

The National Electrical Code gives specific tables for calculating ampacities and conduit sizes for different cables and various conditions.

In this chapter, conduit and cable determinations have been illustrated, but there are other raceway applications. Wireways and trays are used as a raceway to carry the cable. Appropriate derating factors should be applied, based on the National Electrical Code when using other types of raceways.

7

Using Logic To
Simplify Control Systems

Automated process plants are controlled by electrical hardware specified on elementary diagrams and designed by electrical engineers. Standard inexpensive contactors, timing devices, relays and other simple electro-mechanical devices provide control for practically every circuit. In this chapter an approach for analyzing and designing elementary diagrams is developed. Learn the logic language which will form a new way for communicating.

ELECTRICAL HARDWARE

Electrical hardware used for control includes:

• *Relays* — Devices which when energized close. Contacts physically located on the relay will either open or close when the relay is energized.

- *Timers and time delay relays* — Devices whose contacts close or open after a preset time.

- *Pushbuttons* — Devices which are used to actuate a control system; i.e., Stop-Start pushbuttons.

- *Programmers* — Devices whose contacts open and close in a preset sequence.

Present trend is toward plug-in type relays, prepackaged solid-state components; i.e., Square "D" Norpak and mini-computers.

SYMBOLS

Symbols commonly used in electrical control schematics are illustrated in Figure 7-1. These symbols are based on the Joint Industrial Council (JIC) Standards.

THE ELECTRICAL ELEMENTARY (SCHEMATIC)

Figure 7-2 illustrates a typical electrical elementary diagram. Notice that line identifications on the left are used as references to locate relay contacts on the right. For example, relay R_1 has a normally open contact on line six.

STEPS FOR ANALYZING ELECTRICAL CONTROL CIRCUITS

- Look at one line of operation at a time.

- Trace a path of power from left to right. Every contact to the left of the electrically operated line must be closed for the device to operate.

- Each line is identified with a consecutive number on the left.

- Numbers at the right of the line next to a relay show on what lines the device has contacts. Underlined numbers indicate a normally closed contact.

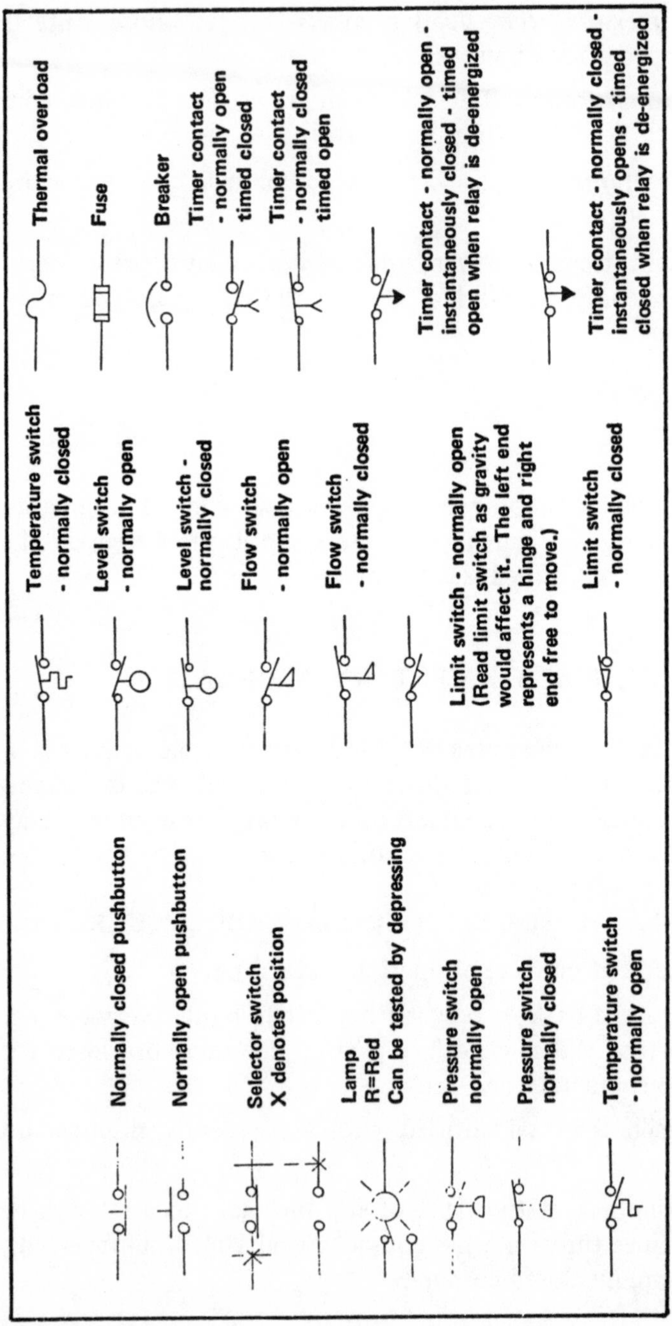

Normally closed pushbutton

Normally open pushbutton

Selector switch
X denotes position

Lamp
R=Red
Can be tested by depressing

Pressure switch
normally open

Pressure switch
normally closed

Temperature switch
- normally open

Temperature switch
- normally closed

Level switch
- normally open

Level switch -
normally closed

Flow switch
- normally open

Flow switch
- normally closed

Limit switch - normally open
(Read limit switch as gravity
would affect it. The left end
represents a hinge and right
end free to move.)

Limit switch
- normally closed

Thermal overload

Fuse

Breaker

Timer contact
- normally open
timed closed

Timer contact
- normally closed
timed open

Timer contact - normally open -
instantaneously closed - timed
open when relay is de-energized

Timer contact - normally closed -
instantaneously opens - timed
closed when relay is de-energized

Figure 7-1. Electrical Symbols

Simple Rules

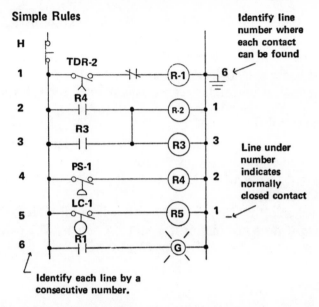

Identify each line by a
consecutive number.

Figure 7-2 Format of an Electrical Elementary Diagram

SIMPLE CONTROL SCHEMES

In order to control a motor, a starter is required. A typical motor elementary diagram, including the power portion, is illustrated in Figure 7-3. HCA represents the holding coil or contactor of the starter. HCA is simply a relay which has contacts capable of interrupting power to the associated motor.

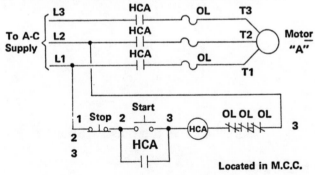

Stop Start - Full Voltage
Non-reversing Motor

Figure 7-3. Elementary Diagram for a FVNR Motor

If the stop-start pushbutton were located locally, 3 wires, numbers 1, 2, 3 would need to run from the field to the M.C.C.

The control voltage in Figure 7-3 is 480 volts, but if the circuit had a solenoid valve, limit switch, etc., 110 volts would be required. (Control devices are usually rated for 110 volts.)

INTERLOCKING

Two typical interlocking schemes are illustrated in Figure 7-4. Scheme "A" illustrates the case where Motor "B" can not be started unless Motor "A" is running. If Motor "A" stops so will Motor "B".

Scheme "B" illustrates the case where once Motor "B" is running it does not matter if Motor "A" stops. The permissive interlock is only required to start Motor "B".

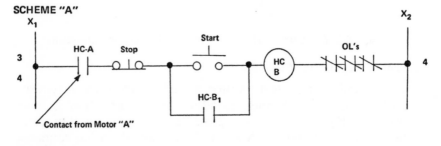

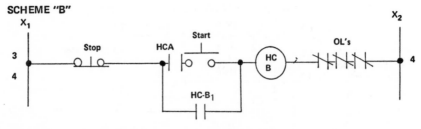

Figure 7-4. Two Interlocking Schemes

REVERSING MOTORS

To change the rotation of a motor it is necessary to switch any two of the motor leads. Figure 7-5 illustrates the elementary for a reversing motor.

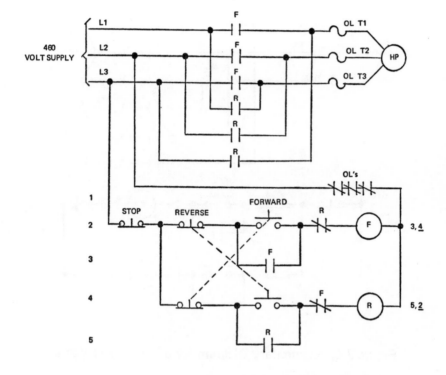

Figure 7-5. Elementary Diagram for a FVR Motor

TWO-SPEED MOTORS

To change the speed of the motor requires changing the effective number of poles. This is accomplished either by using a motor with two separate windings or a motor whose windings are taped so that they can be connected in two ways. Six power leads are required to change the windings. Figure 7-6 illustrates an elementary diagram for a two-speed motor with separate windings. The control portion of the elementary is similar to that of a reversing motor (Figure 7-5) except that six overloads are required.

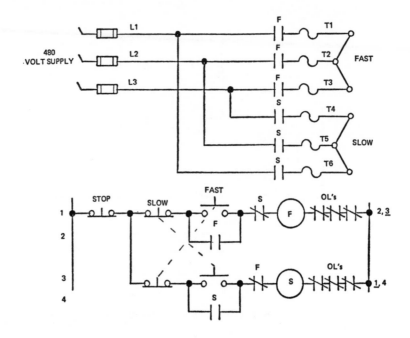

Figure 7-6. Elementary Diagram for a Two-Speed Motor

DESCRIPTION OF OPERATION
AND LOGIC DIAGRAMS

To describe how a process operates it is necessary to establish a logic diagram or description of operation. From this description the electrical schematic is designed.

SIM 7-1

From the description of operations, draw an electrical schematic.

The HVAC design is as follows:

(a) When exhaust fan #1 is operated, damper EP valve is energized (electric to pneumatic).

(b) PE (pneumatic to electric) switch prevents #2 fan from starting.

Answer

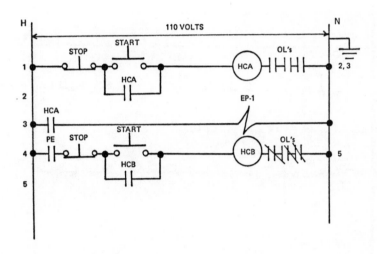

SIM 7-2

Draw elementaries for each scheme.

(a) When Level Switch LCL #1 in Tank "A" reaches "high" level, Pump #3 is started. Pump will run for ten minutes before it will turn off automatically.

(b) Pump #4 which feeds Tank "A" can be started manually and will automatically stop when high level occurs (LCHA).

(c) There are three valves which are used at a manifold station. When Valve "A" is energized Valve "C" can not be energized. When Valve "B" is energized, Valves "A" and "C" can not be energized. Once Valve "C" is energized, Valves "A" and "B" have no effect.

Answer

It should be noted that several electrical schematics may all be correct but look different. Another point is that several people may interpret the above descriptions differently. One possible solution is illustrated.

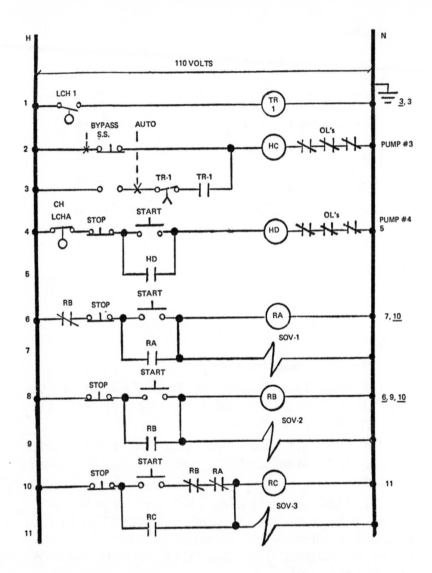

LOGIC DIAGRAMS AND MINI-COMPUTERS

Logic diagrams use symbols to describe the operation whereas the elementary diagram uses words. The logic diagram is an effective tool to convey process requirements since it is understood by all engineers. To describe a process, the logic diagram uses symbols for the three words "And," "Or," "Not." Since any logic can be conveyed with these three words it offers a simple, exact means to describe process requirements. Figure 7-7 illustrates symbols commonly used in logic. These symbols are combined to describe a process. It is possible to design a complete electrical control scheme by using a mini-computer and a logic diagram. A mini-computer is basically a packaged combination of memory units which can be easily programmed for any process. Unlike relays they are not hard-wired, so that changes in operation can be readily made by just reprogramming the unit. It is possible to use the input directly from the logic diagram to program the mini-computer without drawing the elementary diagram as an intermediate step.

Thus logic diagrams can save design time when used with new programmable devices. They are still useful in the development of conventional control schemes since they offer a visual tool which can be understood not only by the electrical engineer, but by the process and mechanical engineer as well. This means that prior to the start of the electrical elementaries and interconnecting drawings the process can be resolved. This in itself can minimize costly design changes.

HOW TO ANALYZE LOGIC DIAGRAMS

To analyze a logic diagram it is necessary to determine the various inputs required to actuate the logic gate. When time delays are incorporated into the design it is necessary to determine the "state" of the process at various time periods. One method to accomplish this is to place a "1" or a "0" after each logic symbol.

A "1" indicates a signal present and a "0" indicates the absence of a signal. Always try to establish the initial state of the circuit. The logic elements can be combined in any order to describe the electrical circuit.

Logic Functions	NEMA Symbol	Description
AND		A device which produces an output only when every input is present—represents contacts in series.
OR		A device which produces an output when one input (or more) is present—represents contacts in parallel.
NOT		A device which produces an output only when the input is absent — represents a normally closed contact.
ON DELAY TIMER		A device which produces an output following a definite time delay after its input is applied.
OFF DELAY TIMER		A device whose output is removed following a definite time delay after its input is removed.

Figure 7-7. Logic Symbols

SIM 7-3

SWITCH A

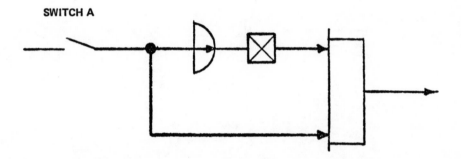

What will happen when Switch "A" is closed?

Draw an elementary.

Answer

Step 1   Identify *initial* operating and final state of logic.

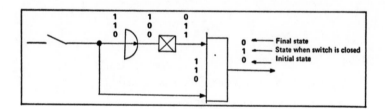

Comments—The above logic shows how various elements can be combined. Note that two signals from switch "A" are used. One goes through a "Time Delay" and "Not" gate and the other is an instantaneous signal fed directly into an "And" gate. Both signals are required to give a pulse after the switch is closed.

Step 2   Look at output — a pulse when equipment is activated.

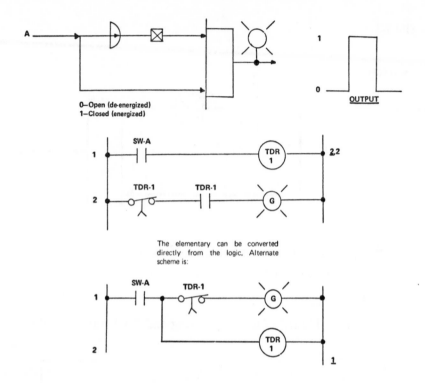

0—Open (de-energized)
1—Closed (energized)

OUTPUT

The elementary can be converted directly from the logic. Alternate scheme is:

SIM 7-4

What will happen when Sw. 1 is closed?
Draw an elementary diagram.

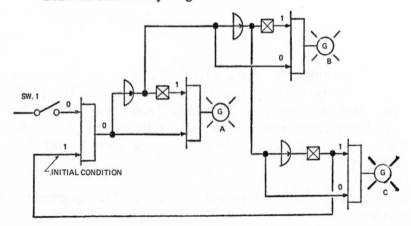

SW. 1

INITIAL CONDITION

Answer

How to Analyze Logic

(a)   Determine initial condition with switch open.

(b)   By placing "1s" and "0s" on above diagram determine operation when switch is closed.

(c)   Always check final state to see if logic resets.

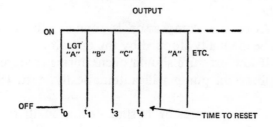

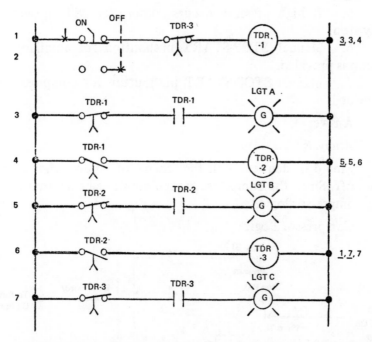

The above elementary and associated logic represent sequential pulses when the switch is closed.

SIM 7-5

From the description of operation, draw a logic diagram and an elementary.

1.   In order for the compressor to operate (C-1), the Oil Pump (P-2) should be running for 5 minutes. An interlock should be provided so that if an operator pressed the start button, he must keep the button depressed until the above is accomplished.

2.   After the compressor is running, the auxiliary oil pump will be manually stopped by the operator.

3.   If low pressure should occur during compressor operation, auxiliary oil pump will automatically start. (Stop same as 1.)

4.   When compressor is shut down, auxiliary oil pump will automatically come on and automatically stop after 5 minutes.

5.   If high pressure occurs compressor will automatically stop.

6.   Manual STOP-START pushbutton for auxiliary oil pump is provided.

7.   Manual STOP-START pushbutton for compressor is provided.

Answer

Analysis

Step 4 is identical with the scheme of a pulse when signal goes off. Since the compressor is the simpler scheme, first draw the compressor logic.

Compressor Logic

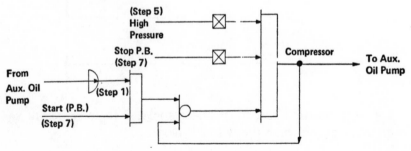

In the auxiliary oil pump scheme the motor can be manually started or through low pressure. Since Step 4 also automatically stops the motor, this step must be drawn independently of the seal in contact.

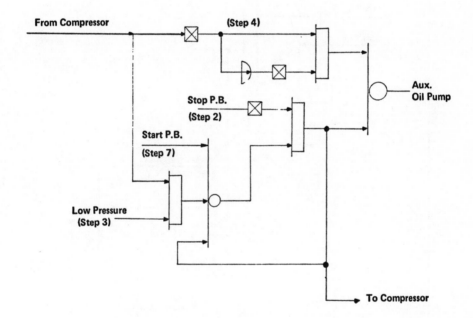

The two basic schemes can be combined, as illustrated on the following page.

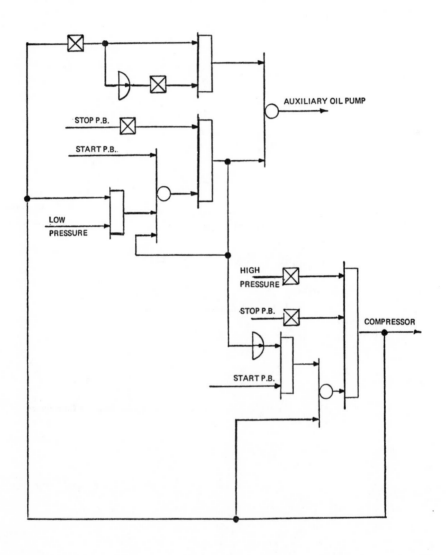

Development of Elementary

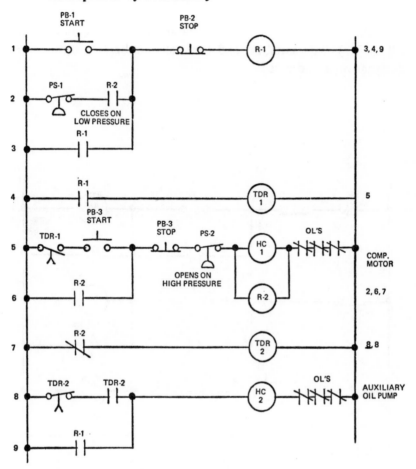

INTERCONNECTION DIAGRAMS

To show the electrician how to wire from an elementary diagram, a physical arrangement referred to as an interconnection diagram is drawn. Vendor's prints showing how the terminal blocks are arranged are used. Each wire on the elementary is assigned a wire number. All devices are connected internally to the terminal blocks for each panel. The electrician needs only to connect the terminals together with the control cable.

SIM 7-6

Draw an interconnection diagram for the elementary in SIM 7-2(c). Wire numbers and terminal numbers have the same designation as indicated.

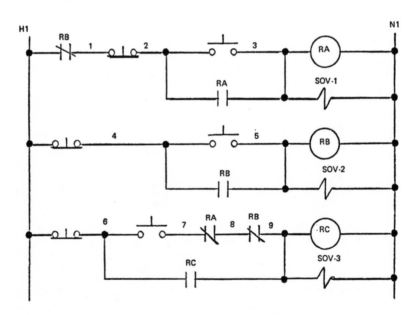

Pushbuttons are located on Pushbutton Panel No. 1. Relays are located on Relay Panel No. 1. Solenoid valves are located locally; 110 volts source is from M.C.C. No. 1.

Answer

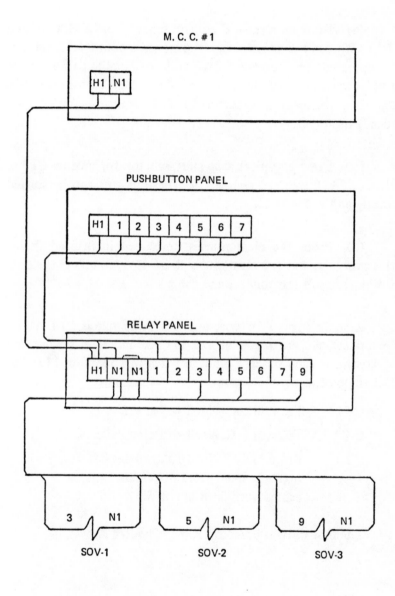

JOB SIMULATION – SUMMARY PROBLEM

JOB 6

(a)   Draw an electrical control scheme for CF-3, C-16 and TP-5. When CF-3 is reversing, the transfer pump TP-5 can not be operated. The conveyor C-16 will not be able to operate in the forward cycle of the centrifuge CF-3. Pushbuttons for C-16 and CF-3 are located on Panel No. 1. Pushbuttons for TP-5 are locally mounted.

(b)   Draw a typical stop-start scheme for motors CTP-6, HF-10, HF-11, BC-13 and SC-19. These motors can be started locally and at the M.C.C.

(c)   From the elementaries developed in (a) and (b) of this problem, and the elementary illustrated, complete the control portions of the conduit and cable schedule.

Assume Type THW wire and minimum size is #14. All interconnections are made at the M.C.C. and field devices are located near each other. Common wires will be jumped locally. Designate conduits as follows:

C-P1   From M.C.C. No. 1 to Panel No. 1.

C-R1   From M.C.C. No. 1 to Relay Panel No. 1.

C-L1   From M.C.C. No. 1 to local devices
          (exclude local pushbuttons since those
          cables were sized in Job 5).

The motor list is based on Job 1, Chapter 2.

Illustrated elementary [see (c)] for motors AG-1, FP-4, CT-9, UH-12 and RD-22.

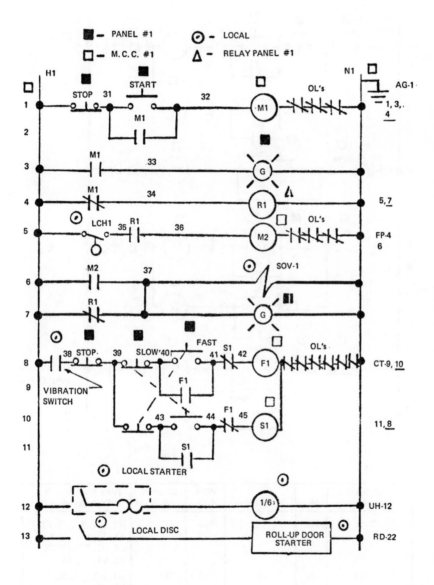

Analysis

(a) and (b)

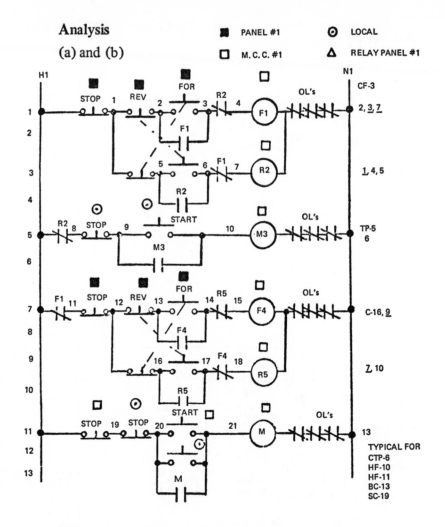

(c)

Conduit Designation	No.	Conductors	Wire Size	Conduit Size	Wire Numbers in Each Conduit
C-P1	20	1/C	#14	1¼"	H1, N1, 2, 3, 5, 6, 11, 13, 14, 16, 17, 31, 32, 33, 37, 38, 40, 41, 43, 44
C-R1	6	1/C	#14	¾"	H1, N1, 34, 35, 36, 37
C-L1	5	1/C	#14	¾"	H1, N1, 35, 37, 38

SUMMARY

Many times it is difficult to obtain a description of operation from the process, project or mechanical engineer. In many plants the elementary diagram is the only document which describes how a process works. Usually only the electrical engineer understands the elementary diagram so it is difficult to insure that the elementary diagram meets the functional requirements. The verbal description which is the input to the elementary diagram can be easily misinterpreted. Since all disciplines understand the simple logic elements, logic diagrams can be initiated and agreed upon. The electrical engineer can then use the logic diagram as the basis for his design. By representing the circuit in terms of a logic diagram the electrical engineer can use Boolean algebra or other switching circuit techniques to simplify the elementary control. Also state-of-the-art techniques using solid-state logic elements or programmable controllers can be specified directly from the logic diagram.

The logic elements form a common language permitting the process, mechanical, project and electrical engineers to create a complex elementary control scheme. The individual who understands the fundamentals of the logic elements can easily apply this important tool in developing complex industrial and power elementary control schemes.

8

Applying Programmable Controllers and Electronic Instrumentation

The basic requirement for developing electrical schematics is to understand the process. The logic diagram is of particular use in describing a process when the activating input is related to an on-off signal. Traditionally, the hardware used to accomplish this logic has been a combination of heavy-duty machine tool relays. A growing trend is the use of solid-state programmable controllers, especially in applications where the process steps repeat continuously. In addition, the current trend is to control process variables, i.e., temperature, pressure, flow level by electronic instrumentation loops. Thus it becomes increasingly important to understand how programmable controllers and instrumentation are applied.

PROGRAMMABLE CONTROLLER

The programmable controller developed out of the needs of the automotive industry. The industry required a control unit which could easily be changed in the plant, easily be maintained and repaired, highly reliable, small, capable to output data to a central data collection system, and competitive in cost to relay and solid-state panels. Out of these requirements developed the first series of programmable controllers. These units were initially designed to accept 115-volt ac inputs, to output 115 volts ac —2 ampere signals capable of actuating solenoid valves, motors, etc. The memory was capable of expansion to a minimum of 4000 words.

Today the programmable controller is far more flexible and reliable than the earlier generations and its use has had far-reaching implications to most industrial plants. Let's examine some of the features of most models.

The basic unit described in Figure 8-1 consists of:
- *Input module*
- *Memory*
- *Processor*
- *Output module*
- *Programming auxiliary*

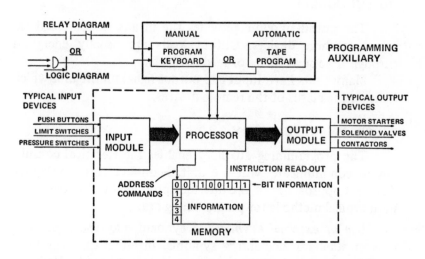

Figure 8-1. Components of a Programmable Controller

INPUT MODULE

The input module usually accepts ac or dc signals from remote devices such as pushbuttons and switches. These signals are then converted to dc levels, filtered and attenuated for use by the processor.

PROCESSOR

This module is the working portion of the programmable controller. All input and outputs are continuously monitored. The status of the input is compared against an established program and instructions are executed to the various outputs. The control function of the processor identifies the memory core locations to be addressed. An internal timing device determines the required sequence to fetch and execute instructions.

MEMORY

The information stored in the memory relates to how the input-output data should be processed. Information is usually stored on magnetic cores.

OUTPUT MODULE

The output module provides the means to command external machine devices. Output loads are usually energized through triac ac switches or reed contacts.

Many units have input and output lights on the unit which indicate the status of the remote devices.

PROGRAMMING AUXILIARY

The programming auxiliary enables the electrical consultant to communicate with the unit. Manufacturers use slightly different approaches to the interface problem but the two common typical methods for programming are:

Use of external keyboard. Keyboard may use logic symbols or standard elementary symbols. In either case, the logic or elementary is directly programmed into the unit. Refer to Figure 8-2.

Use of a Tape. A program is generated by use of a teletypewriter on a computer time-sharing facility. The tape generated can then be loaded into a tape cassette or loader.

Some units offer keyboards with visual displays. The visual displays are useful when input data needs to be verified and when changes occur in the elementary.

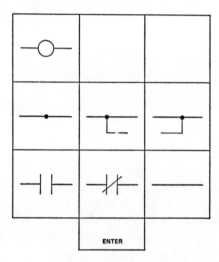

Figure 8-2. Typical Keyboard for Programmable Controller

Another feature of programming is in elementary simplification and simulation. If a tape is used for programming, the information on the tape can be used in a standard elementary simplification program. Any redundancies in the logic can be quickly found. The tape can also be used in conjunction with a simulation program. For any input, the corresponding output can be determined without the actual output being energized. To change sequence of operation, the memory is simply reprogrammed.

CAPABILITIES

The programmable controller is capable of performing the function of hardwired solid-state systems or relay systems.

The basic functions are:

- *Logic Gates*—For a review of logic, see Chapter 7.
- *Timers*—Either on delay or off delay timers. Typical range .1 to 99.9 seconds.
- *Counters*—Used to count input status changes at a rate of 20 per second. Counter size 999 counts.
- *Latches*—Simulates electro-mechanical relay.

- *Shift Registers*—Provides the ability to simultaneously remember the state of several pieces as they move through the manufacturing process.
- *Maintenance*—Many units feature plug-in modules and a preprogrammed diagnostic tape so that problems can be quickly identified and corrected.

APPLICATIONS

There are many programmable controllers available. Each unit may have slightly different features but they all offer the versatility of a non-hardwired controller. The size of the input-output and logic capabilities also vary, but where repeat operations are present, the programmable controller can economically compete with their relay counterparts.

In the case where the units are economically competitive, the programmable controller should seriously be considered since it offers:

- Complete flexibility to modify or expand controls when production needs change.

- A highly reliable solid state control which minimizes downtime.

- The possible reduction of engineering design time (no need to check back of panel wiring, etc.).

- A system compatible with computer simplification and simulation techniques.

- Fast diagnostics and maintenance.

INSTRUMENTATION

Modern Process Control hinges around electronic instrumentation. For years pneumatic instrumentation dominated the market, but today most control loops have been replaced by their electronic counterparts.

CLOSED LOOP

To control a process requires a closed loop system. In the example of controlling level, if the level is too high the level is reduced. Figure 8-3 shows a typical closed loop system.

Where:

C   is the controlled variable
R   is the reference or set point
E   is the error or deviation
M   is the variable manipulated by the controller.

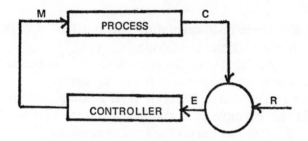

Figure 8-3. Typical Closed Loop System

Figure 8-4 shows a typical process control diagram illustrating level control.

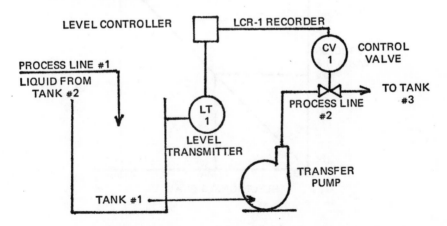

Figure 8-4. Typical Level Control Process

In this process level transmitter (LT-1) sends the electronic signal to level controller recorder (LCR-1). If the level is too low, control valve CV-1 is throttled. On the other hand, if the level is too high CV-1 is opened. Thus, the process is controlled. This example illustrates the fundamental elements of a control loop, namely:

Input (transmitter)
Controller
Output (control valve).

SIGNAL

In electronic instrumentation the standard signal used to convey process variables is 4-20 ma dc.

Figure 8-5 illustrates the process variable in percentage as a function of the corresponding electronic signal.

This means that a full-scale level reading corresponds to 100%. In this example an 8 ma dc signal from LT-1 would correspond to a 25% reading on the level controller.

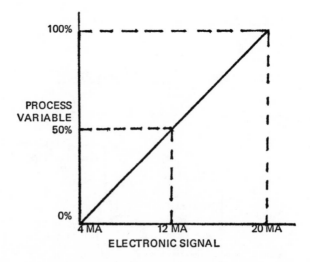

Figure 8-5. Process Variable vs. Electronic Signal

CONTROL LOOP

The basic control circuitry for electronic instrumentation is illustrated in Figure 8-6. This figure illustrates a typical series circuit for the process of Figure 8-4. Each receiving element contains an input impedance.

In a series network, the impedance of each element must be carefully matched.

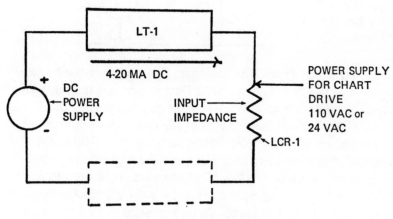

Figure 8-6. Control Loop

TYPES OF CONTROL LOOPS

The several common types of control loops are:

- *Level*
- *Flow*
- *Pressure*
- *Temperature*

Level: As indicated in Figure 8-4, level is controlled by CV-1. Level could also be controlled by a control valve on process line No. 1. Two common types of level transmitters are the float type and differential pressure type.

Flow: As indicated in Figure 8-7 flow is essentially controlled the way level is controlled The signal from FT-1 is sent to FC-1 and control valve CV-1 is either opened or closed. Two common types of flow transmitters are the magnetic flow meter and orifice meter type.

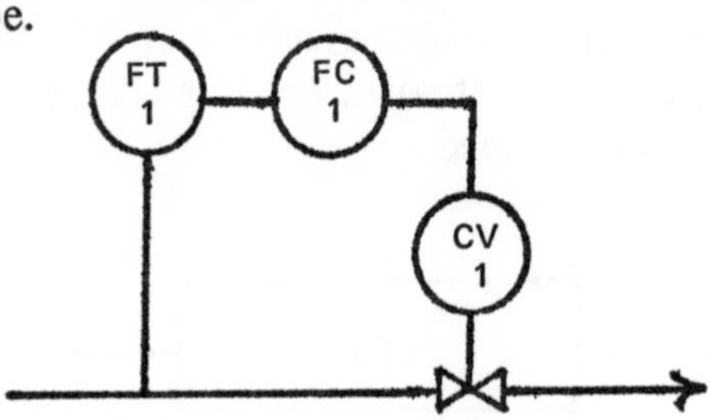

**Figure 8-7.
Flow Control**

Magmeter Type: The magnetic flow meter is the more expensive of the two. The magnetic flow meter uses the same principle of operation as a tachometer or generator. As indicated in Figure 8-8 the fluid flowing acts as the conductor while the pipe is located in the magnetic field caused by the field coils. The electrodes are mounted in a plane at right angles to the magnetic field and act like brushes of a generator. Thus the voltage induced by the moving fluid is brought out by leads for external measurement. The induced voltage can be converted by the transmitter to direct current. For magmeters to work properly, the fluid must conduct electricity and the tube must be liquid full.

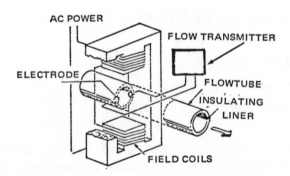

Figure 8-8. Magnetic Flow Meter

Orifice Meter Type: As liquid flows through a restriction in a pipe orifice plate, a change in pressure occurs as illustrated in Figure 8-9. The flow rate is calculated by measuring the differential pressure across the orifice plate.

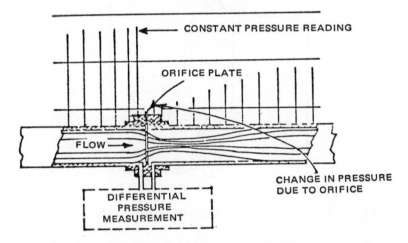

Figure 8-9. Orifice Flow Meter

Pressure: Pressure control is used to maintain a specified pressure or is used as an indirect measurement of level or flow, as indicated previously.

Temperature: The thermocouple is one of the most frequently used methods for measuring temperatures between 500 and 1500°C. Strictly speaking the thermocouple and its associated control do not fall into the typical electronic instrumentation category.

The operation of a thermocouple is based on the principle that an electromotive force (emf) is developed when two different metals come in contact. The emf developed is dependent on the metals involved and the *temperature* of the junctions. The emf signal developed is expressed in millivolts. Due to the nature of the thermocouple, wire splices should be avoided. Two types of measuring circuits are used in conjunction with thermocouples.

Figure 8-10 illustrates a typical thermocouple circuit using a Galvanometer.

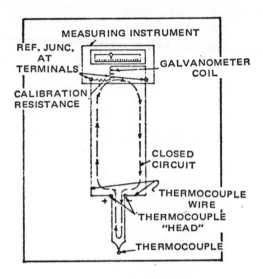

Figure 8-10. Basic Galvanometer Thermocouple Circuit

Figure 8-11 illustrates a typical thermocouple circuit using a potentiometer.

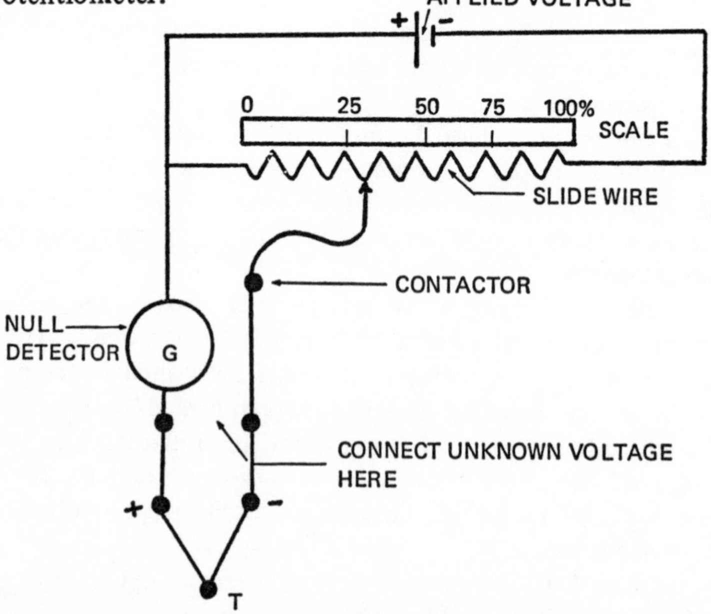

Figure 8-11. Basic Potentiometer Thermocouple Circuit

WIRING METHODS

The type of cable, and installation of instrument signals is based on the reduction of noise. In general instrument cables are routed away from noise sources such as power cables, motors, generators, etc. Twisted control cables are used to reduce magnetic noise pickup from a nearby noise source.

Table 8-1 summarizes instrumentation wiring methods.

Table 8-1. Instrumentation Wiring Methods

TRANSMITTER TYPE	RECEIVING ELEMENT DESCRIPTION	
	Torque type	Bridge or potentiometer type
	current converted directly to a torque to move a chart recorder or pointer, i.e., pyrometer, Galvanometer	input signal is compared with a standard voltage and amplified to drive a chart or recorder to a balanced or null position
THERMOCOUPLE	twisted pair nonshielded	twisted pair shielded
	NOTE: lead wire must be the same material as the thermocouple, i.e., iron constantan	
MAGNETIC FLOW METERS		twisted pair shielded
PNEUMATIC CURRENT TRANSDUCERS, DIFFERENTIAL PRESSURE FLOW METERS	twisted pair nonshielded	*twisted pair nonshielded *NOTE: for a high noise environment use twisted pair shielded.

9

Utilizing Intrinsic Safety in Hazardous Areas

Electrical equipment must be installed in accordance with the area classification. The National Electrical Code (NEC), Articles 500-503, is the basic reference used to define hazardous locations and electrical installation requirements. The NEC has been the recommended practice of both the American Petroleum Institute (API) and the Instrument Society of America (ISA).

The emphasis in the NEC has been on power and lighting applications in hazardous areas. The ISA standards have supplemented NEC for instrument installations in hazardous areas. In the last decade instrument systems have tended towards small-case electrical control systems which are fast and versatile.

The increase in electrical instrumentation, the economic motivation for safer and more efficient installations has led to the development of electrical control systems which are intrinsically safe. This chapter will review the fundamentals of area classifications, methods for locating electrical equipment in hazardous areas, and applications of intrinsic safety.

AREA CLASSIFICATION

At the beginning of a project the electrical engineer develops an area classification drawing which shows the plant location and the classification of each area. The process requirements, input from the client and from the insurer are the main ingredients in determining the classification to be assigned. The nature and degree of the hazard is specified in terms of its class, its grouping, and its division.

The generic nature of the hazardous material is denoted by its class, as defined in Table 9-1.

Table 9-1. Class Hazard Designations

> • Where flammable gases or vapors may be present in the air in quantities sufficient to produce an explosive or ignitable mixture the location is referred to as Class I.
>
> • Where combustible dusts may be present in the air in quantities sufficient to produce an explosive or ignitable mixture the location is referred to as Class II.
>
> • Where ignitable fibers or flyings may be present in the air in quantities sufficient to produce an ignitable mixture the location is referred to as Class III.

To more specifically describe the nature of the hazard a subclassification or group letter is assigned, as defined in Table 9-2.

Table 9-2. Group Hazard Designations

Group A —	atmospheres containing acetylene.
Group B —	atmospheres containing hydrogen or gases or vapors of equivalent hazard such as manufactured gas, and butadiene, ethylene oxide, propylene oxide.
Group C —	atmospheres containing ethyl ether vapors, ethylene, or cyclopropane, acetaldehyde, isoprene, unsymmetrical dimethylhydrazine.
Group D —	atmospheres containing gasoline, hexane, naptha, benzine, butane, propane, alcohol, acetone, benzol, lacquer solvent vapors, or natural gas, and acrylonitrile, ethylene dichloride, propylene, styrene, vinyl acetate, vinyl chloride, p-xylene.
Group E —	atmospheres containing metal dust including aluminum, magnesium, and their commercial alloys, and other metals of similar hazardous characteristics.
Group F —	atmospheres containing carbon black, coal, or coke dust.
Group G —	atmospheres containing flour, starch or grain dust.

The last designation indicates the degree of hazard due to the concentration of the material present. Table 9-3 summarizes division designations.

Table 9-3. Division Hazard Designation

Division 1 — the location is likely to have hazardous concentrations existing (especially below grade and enclosed areas) due to: continuous, intermittent or periodic discharges under normal operating conditions.

Division 2 — the location is presumed hazardous only under abnormal conditions. Examples include: locations where flammable liquids or gases are handled in closed systems or confined to closed containers, areas adjacent to Division 1 areas, areas which could become hazardous as a result of a failure of a forced ventilation system. Locations containing hazardous materials as indicated and are outdoors are usually classified as Division 2.

UNCLASSIFIED

An unclassified area or nonhazardous location is so designated when the occurrence of a flammable material from an operation or apparatus is so infrequent that it is not necessary to classify it.

Examples include: closed systems including only pipes, valves, fittings, flanges and similar accessories, adequate ventilated areas where flammable materials are contained in suitable well-maintained closed systems. An adequately ventilated area is either substantially open or is artificially ventilated. In practice, areas surrounding permanent ignition sources such as fired equipment are usually unclassified without regard to the probability of release of flammable vapor.

Guidelines to determine the extent of area classifications based on the location of the source and the type of enclosure are contained in API RP500A. Figures 9-1, 9-2, and 9-3 are reproduced from this API publication and illustrate typical area classifications.

ELECTRICAL INSTALLATIONS IN HAZARDOUS AREAS

Four common ways in which electrical arcing devices can be installed in hazardous areas are:

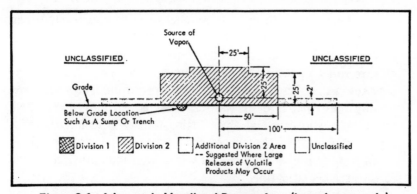

Figure 9-1. Adequately Ventilated Process Area (hazard near grade)

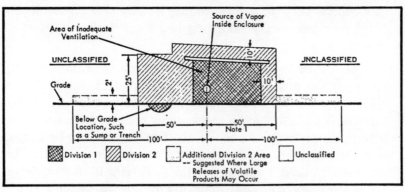

Figure 9-2. Process Area with Inadequate Ventilation

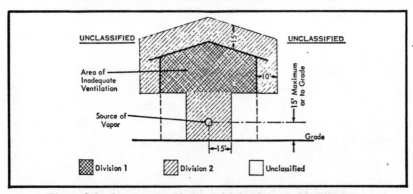

Figure 9-3. Compressor Shelter with Inadequate Ventilation

Reproduced with permission from the API Publication RP-5008—*Recommended Practice for Classification of Areas for Electrical Installations in Petroleum Refineries.* Copies of this publication may be obtained from the American Petroleum Institute, 2101 L Street, N.W., Washington, D.C. 20037, Attn: Publication Sales.

• One way is to provide an enclosure so constructed that if ignition does occur the flame cannot propagate outside the enclosure and spread to the surrounding atmosphere. The enclosure must contain the explosion without danger. These enclosures are referred to as explosion-proof housings and are usually constructed of heavy wall cost construction, threaded or bolted enclosures and close fitting flanges with long narrow gaps through which hot gases can escape. When an explosion-proof-installation is required all wiring methods must conform to the area classification.

• A second method applies to Division 2 areas. In this case general-purpose enclosures may be used if make-break contacts are hermetically sealed against the entrance of gases or vapors.

• A third method is to eliminate the explosion hazard by removing the potential source of ignition. This can be accomplished by providing a purging system to prevent flammable material from entering the enclosure by maintaining a small positive pressure inside the enclosure. The enclosure to be purged must be strong enough to withstand the positive pressure. Windows must be shatterproof, ¼″ thick. Purging of instruments is a common practice. Remember purging of instruments is not acceptable when the instrument itself handles flammable materials.

• A fourth method is to eliminate the explosion hazard by designing a circuit incapable of producing a spark which can ignite any mixture with air of the flammable material concerned. Thus, the equipment or circuit is referred to as "intrinsically safe" and the wiring methods, enclosures, etc. associated with the loop can then conform to that of a nonhazardous classification.

APPLYING INTRINSIC SAFETY

The National Electrical Code allows for use of intrinsic safe systems as stated in Article 500-1.

Equipment and associated wiring approved as intrinsically safe shall be permitted in any hazardous (classified) location for which it is approved, and the provisions of Articles 500 through 517 shall not be considered applicable to such installations. Means shall be provided to prevent the passage of gases and vapors. Intrinsically safe equipment and wiring shall not be capable of releasing sufficient electrical or thermal energy under normal or abnormal conditions to cause ignition of a specific hazardous atmospheric mixture in its most easily ignited concentration.

Abnormal conditions shall include accidental damage to any field installed wiring, failure of electrical components, application of overvoltage, adjustment and maintenance operations, and other similar conditions.

Intrinsically safe equipment is designed and rated as such to insure that energy levels released will not, under any conditions, cause ignition of the surrounding atmosphere. This means that all elements connected in the circuit must be rated as intrinsically safe.

The use of barrier circuitry has greatly simplified application of intrinsic safety. The barrier circuitry permits interfacing of hazardous and nonhazardous areas. The equipment located in the nonhazardous control room need not be intrinsically safe as long as the equipment and the outside loop are isolated through an intrinsically safe barrier. The barrier is located in the nonhazardous location and insures that ignition capable voltages or currents will not reach the hazardous location even if full voltage is applied at the input barrier device terminals. Thus, any equipment can be used in the nonhazardous location as long as it contains no voltage higher than the barrier design rating, typically 250 volts rms. Figure 9-4 illustrates a typical barrier circuit. The barrier consists of a Zener diode network with resistors and is designed to have a negligible effect on the normal instrument signal. The diodes limit the power which can enter the hazardous areas by conducting as the voltage level increases. The barrier is designed to handle the worst possible fault that can occur. The barrier concept has increased the acceptance of intrinsic safe systems.

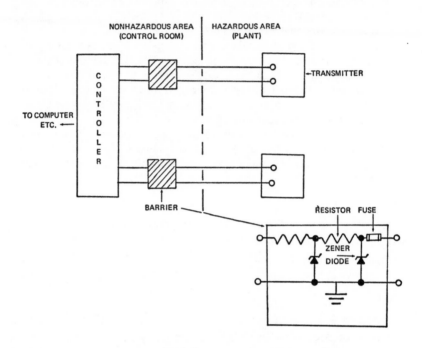

Figure 9-4. Barrier Circuit

INSTALLING
INTRINSICALLY SAFE SYSTEMS

The National Electrical Code does not state what requirements must be met when installing intrinsically safe systems. Guidelines which have developed from ISA Committees include:

• Installation of wiring for intrinsically safe systems in Division 1 areas need only meet the requirements for ordinary locations.

• Circuits which are not intrinsically safe should be isolated from intrinsically safe wiring.

BENEFITS OF
INTRINSICALLY SAFE SYSTEMS

Several benefits are derived by utilizing intrinsically safe systems. The major plus is that explosion-proof junction boxes and conduit are unnecessary. Installation costs are especially reduced in applications involving complex systems and long wiring runs. Another benefit is that maintenance is lower than explosion-proof installations. Removing explosion-proof covers and proper replacement of bolts are eliminated, when intrinsically safe systems are used.

10

Protective Relaying for Power Distribution Systems

Due to possible equipment failure or human error, it is necessary to provide protection devices. These devices minimize system damage and limit the extent and duration of service interruption when failure occurs. The main goal of protection coordination is to isolate the affected portion of the system quickly while at the same time maintaining normal service for the remainder of the system. In other words, the electrical system must provide protection and selectivity to insure that the fault is minimized while other parts of the system not directly involved are held in until other protective devices clear the trouble. Protective devices such as fuses and circuit breakers have time current characteristics which determine the time it takes to clear the fault. In the case of circuit breakers it is possible to adjust the characteristics while fuse characteristics are non-adjustable.

Protective relays are another way of achieving selective coordination and are required to operate power breakers above 600 volts. By definition a protective relay is a device which when energized by suitable currents, voltages, or both, responds to the magnitude and relationships of these signals to indicate or isolate an abnormal operating condition. These relays have adjustable settings and can be used to actuate the opening of circuit breakers under various fault conditions.

In Chapter 4, the concept of the one-line diagram was introduced. In order to complete the power distribution system it is necessary to show on the one-line diagram the protective relaying required and the breakers affected. To properly set the protective devices it is necessary to know the fault currents which occur at various portions of the system. This chapter will illustrate typical applications of protective devices.

THE OVERCURRENT RELAY (Device 51)

Every protective relay has an associated number. Some of the standard designations are listed in Table 10-1. The overcurrent relay is designated as Device 51. This relay is used for overcurrent protection and is current sensitive. The one-line diagram utilizing the overcurrent relay is shown in Figure 10-1.

Table 10-1. Protective Device Numbering and Functions

DEVICE NUMBER	DEFINITION AND FUNCTION
1	master element is the initiating device, such as a control switch, voltage relay, float switch, etc., which serves either directly, or through such permissive devices as protective and time-delay relays to place an equipment in or out of operation.
2	time-delay starting, or closing, relay is a device which functions to give a desired amount of time delay before or after any point or operation in a switching sequence or protective relay system, except as specifically provided by device functions 62 and 79 described later.
3	checking or interlocking relay is a device which operates in response to the position of a number of other devices, or to a number of predetermined conditions in an equipment to allow an operating sequence to proceed, to stop, or to provide a check of the position of these devices or of these conditions for any purpose.
4	master contactor is a device, generally controlled by device No. 1 or equivalent, and the necessary permis-

Table 10-1. Protective Device Numbering and Functions (continued)

DEVICE NUMBER	DEFINITION AND FUNCTION
4 (con't.)	sive and protective devices, which serves to make and break the necessary control circuits to place an equipment into operation under the desired conditions and to take it out of operation under other or abnormal conditions.
5	stopping device functions to place and hold an equipment out of operation.
6	starting circuit breaker is a device whose principal function is to connect a machine to its source of starting voltage.
7	anode circuit breaker is one used in the anode circuits of a power rectifier for the primary purpose of interrupting the rectifier circuit if an arc back should occur.
8	control power disconnecting device is a disconnecting device—such as a knife switch, circuit breaker or pull-out fuse block—used for the purpose of connecting and disconnecting, respectively, the source of control power to and from the control bus or equipment. note: Control power is considered to include auxiliary power which supplies such apparatus as small motors and heaters.
9	reversing device is used for the purpose of reversing a machine field or for performing any other reversing functions.
10	unit sequence switch is used to change the sequence in which units may be placed in and out of service in multiple-unit equipments.
11	Reserved for future application.
12	over-speed device is usually a direct-connected speed switch which functions on machine overspeed.
13	synchronous-speed device, such as a centrifugal speed switch, a slip-frequency relay, a voltage relay, an undercurrent relay or any type of device, operates at approximately synchronous speed of a machine.
14	under-speed device functions when the speed of a machine falls below a predetermined value.

/more/

Table 10-1. Protective Device Numbering and Functions (continued)

DEVICE NUMBER	DEFINITION AND FUNCTION
15	speed or frequency, matching device functions to match and hold the speed or the frequency of a machine or of a system equal to, or approximately equal to, that of another machine, source or system.
16	Reserved for future application.
17	shunting, or discharge, switch serves to open or to close a shunting circuit around any piece of apparatus (except a resistor), such as a machine field, a machine armature, a capacitor or a reactor. note: This excludes devices which perform such shunting operations as may be necessary in the process of starting a machine by devices 6 or 42, or their equivalent, and also excludes device 73 function which serves for the switching of resistors.
18	accelerating or decelerating device is used to close or cause the closing of circuits which are used to increase or to decrease the speed of a machine.
19	starting-to-running transition contactor is a device which operates to initiate or cause the automatic transfer of a machine from the starting to the running power connection.
20	electrically operated valve is a solenoid- or motor-operated valve which is used in a vacuum, air, gas, oil, water, or similar, lines. note: The function of the valve may be indicated by the insertion of descriptive words such as "Brake" or "Pressure Reducing" in the function name, such as "Electrically Operated Brake Valve."
21	distance relay is a device which functions when the circuit admittance, impedance or reactance increases or decreases beyond predetermined limits.
22	equalizer circuit breaker is a breaker which serves to control or to make and break the equalizer or the current-balancing connections for a machine field, or for regulating equipment, in a multiple-unit installation.
23	temperature control device functions to raise or to lower the temperature of a machine or other apparatus,

/more/

Table 10-1. Protective Device Numbering and Functions
(continued)

DEVICE NUMBER	DEFINITION AND FUNCTION
23 (con't.)	or of any medium, when its temperature falls below, or rises above, a predetermined value. note: An example is a thermostat which switches on a space heater in a switchgear assembly when the temperature falls to a desired value as distinguished from a device which is used to provide automatic temperature regulation between close limits and would be designated as 90T.
24	Reserved for future application.
25	synchronizing, or synchronism-check, device operates when two a-c circuits are within the desired limits of frequency, phase angle or voltage, to permit or to cause the paralleling of these two circuits.
26	apparatus thermal device functions when the temperature of the shunt field or the armortisseur winding of a machine, or that of a load limiting or load shifting resistor or of a liquid or other medium exceeds a predetermined value; or if the temperature of the protected apparatus, such as a power rectifier, or of any medium decreases below a predetermined value.
27	undervoltage relay is a device which functions on a given value of undervoltage.
28	Reserved for future application.
29	isolating contactor is used expressly for disconnecting one circuit from another for the purposes of emergency operation, maintenance, or test.
30	annunciator relay is a nonautomatically reset device which gives a number of separate visual indications upon the functioning of protective devices, and which may also be arranged to perform a lockout function.
31	separate excitation device connects a circuit such as the shunt field of a synchronous converter to a source of separate excitation during the starting sequence; or one which energizes the excitation and ignition circuits of a power rectifier.
32	directional power relay is one which functions on a desired value of power flow in a given direction, or upon reverse power resulting from arc back in the anode or cathode circuits of a power rectifier.

/more/

Table 10-1. Protective Device Numbering and Functions (continued)

DEVICE NUMBER	DEFINITION AND FUNCTION
33	**position switch** makes or breaks contact when the main device or piece of apparatus, which has no device function number, reaches a given position.
34	**motor-operated sequence switch** is a multi-contact switch which fixes the operating sequence of the major devices during starting and stopping, or during other sequential switching operations.
35	**brush-operating, or slip-ring short-circuiting, device** is used for raising, lowering, or shifting the brushes of a machine, or for short-circuiting its slip rings, or for engaging or disengaging the contacts of a mechanical rectifier.
36	**polarity device** operates or permits the operation of another device on a predetermined polarity only.
37	**undercurrent or underpower relay** is a device which functions when the current or power flow decreases below a predetermined value.
38	**bearing protective device** is one which functions on excessive bearing temperature, or on other abnormal mechanical conditions, such as undue wear, which may eventually result in excessive bearing temperature.
39	Reserved for future application.
40	**field relay** is a device that functions on a given or abnormally low value or failure of machine field current, or on an excessive value of the reactive component of armature current in an a-c machine indicating abnormally low field excitation.
41	**field circuit breaker** is a device which functions to apply, or to remove, the field excitation of a machine.
42	**running circuit breaker** is a device whose principal function is to connect a machine to its source of running voltage after having been brought up to the desired speed on the starting connection.
43	**manual transfer or selector device** transfers the control circuits so as to modify the plan of operation of the switching equipment or of some of the devices.

/more/

Table 10-1. Protective Device Numbering and Functions
(continued)

DEVICE NUMBER	DEFINITION AND FUNCTION
44	unit sequence starting relay is a device which functions to start the next available unit in a multiple-unit equipment on the failure or on the non-availability of the normally preceding unit.
45	Reserved for future application.
46	reverse-phase, or phase-balance, current relay is a device which functions when the polyphase currents are of reverse-phase sequence, or when the polyphase currents are unbalanced or contain negative phase-sequence components above a given amount.
47	phase-sequence voltage relay is a device which functions upon a predetermined value of polyphase voltage in the desired phase sequence.
48	incomplete sequence relay is a device which returns the equipment to the normal, or off, position and locks it out if the normal starting, operating or stopping sequence is not properly completed within a predetermined time.
49	machine, or transformer, thermal relay is a device which functions when the temperature of an a-c machine armature, or of the armature or other load carrying winding or element of a d-c machine, or converter or power rectifier or power transformer (including a power rectifier transformer) exceeds a predetermined value.
50	instantaneous overcurrent, or rate-of-rise relay is a device which functions instantaneously on an excessive value of current, or on an excessive rate of current rise, thus indicating a fault in the apparatus or circuit being protected.
51	a-c time overcurrent relay is a device with either a definite or inverse time characteristic which functions when the current in an a-c circuit exceeds a predetermined value.
52	a-c circuit breaker is a device which is used to close and interrupt an a-c power circuit under normal conditions or to interrupt this circuit under fault or emergency conditions.

/more/

Table 10-1. Protective Device Numbering and Functions (continued)

DEVICE NUMBER	DEFINITION AND FUNCTION
53	**exciter or d-c generator relay** is a device which forces the d-c machine field excitation to build up during starting or which functions when the machine voltage has built up to a given value.
54	**high-speed d-c circuit breaker** is a circuit breaker which starts to reduce the current in the main circuit in 0.01 second or less, after the occurrence of the d-c overcurrent or the excessive rate of current rise.
55	**power factor relay** is a device which operates when the power factor in an a-c circuit becomes above or below a predetermined value.
56	**field application relay** is a device which automatically controls the application of the field excitation to an a-c motor at some predetermined point in the slip cycle.
57	**short-circuiting or grounding device** is a power or stored energy operated device which functions to short-circuit or to ground a circuit in response to automatic or manual means.
58	**power rectifier misfire relay** is a device which functions if one or more of the power rectifier anodes fails to fire.
59	**overvoltage relay** is a device which functions on a given value of overvoltage.
60	**voltage balance relay** is a device which operates on a given difference in voltage between two circuits.
61	**current balance relay** is a device which operates on a given difference in current input or output of two circuits.
62	**time-delay stopping, or opening, relay** is a time-delay device which serves in conjunction with the device which initiates the shutdown, stopping or opening operation in an automatic sequence.
63	**liquid or gas pressure, level, or flow relay** is a device which operates on given values of liquid or gas pressure, flow or level, or on a given rate of change of these values.

/more/

Table 10-1. Protective Device Numbering and Functions
(continued)

DEVICE NUMBER	DEFINITION AND FUNCTION
64	**ground protective relay** is a device which functions on failure of the insulation of a machine, transformer or of other apparatus to ground, or on flashover of a d-c machine to ground. **note:** This function is assigned only to a relay which detects the flow of current from the frame of a machine or enclosing case or structure of a piece of apparatus to ground, or detects a ground on a normally ungrounded winding or circuit. It is not applied to a device connected in the secondary circuit or secondary neutral of a current transformer, or current transformers, connected in the power circuit of a normally grounded system.
65	**governor** is the equipment which controls the gate or valve opening of a prime mover.
66	**notching, or jogging, device** functions to allow only a specified number of operations of a given device, or equipment, or a specified number of successive operations within a given time of each other. It also functions to energize a circuit periodically, or which is used to permit intermittent acceleration or jogging of a machine at low speeds for mechanical positioning.
67	**a-c directional overcurrent relay** is a device which functions on a desired value of a-c overcurrent flowing in a predetermined direction.
68	**blocking relay** is a device which initiates a pilot signal for blocking of tripping on external faults in a transmission line or in other apparatus under predetermined conditions, or co-operates with other devices to block tripping or to block reclosing on an out-of-step condition or on power swings.
69	**permissive control device** is generally a two-position, manually operated switch which in one position permits the closing of a circuit breaker, or the placing of an equipment into operation, and in the other position prevents the circuit breaker or the equipment from being operated.
70	**electrically operated rheostat** is a rheostat which is used to vary the resistance of a circuit in response to some means of electrical control.
71	Reserved for future application. */more/*

Table 10-1. Protective Device Numbering and Functions (continued)

DEVICE NUMBER	DEFINITION AND FUNCTION
72	**d-c circuit breaker** is used to close and interrupt a d-c power circuit under normal conditions or to interrupt this circuit under fault or emergency conditions.
73	**load-resistor contactor** is used to shunt or insert a step of load limiting, shifting, or indicating resistance in a power circuit, or to switch a space heater in circuit, or to switch a light, or regenerative, load resistor of a power rectifier or other machine in and out of circuit.
74	**alarm relay** is a device other than an annunciator, as covered under device No. 30, which is used to operate, or to operate in connection with, a visual or audible alarm.
75	**position changing mechanism** is the mechanism which is used for moving a removable circuit breaker unit to and from the connected, disconnected, and test positions.
76	**d-c overcurrent relay** is a device which functions when the current in a d-c circuit exceeds a given value.
77	**pulse transmitter** is used to generate and transmit pulses over a telemetering or pilot-wire circuit to the remote indicating or receiving device.
78	**phase angle measuring, or out-of-step protective relay** is a device which functions at a predetermined phase angle between two voltages or between two currents or between voltage and current.
79	**a-c reclosing relay** is a device which controls the automatic reclosing and locking out of an a-c circuit interrupter.
80	Reserved for future application.
81	**frequency relay** is a device which functions on a predetermined value of frequency—either under or over or on normal system frequency—or rate of change of frequency.
82	**d-c reclosing relay** is a device which controls the automatic closing and reclosing of a d-c circuit interrupter, generally in response to load circuit conditions.
83	**automatic selective control, or transfer, relay** is a device which operates to select automatically between

/more/

Table 10-1. Protective Device Numbering and Functions
(continued)

DEVICE NUMBER	DEFINITION AND FUNCTION
83 (con't.)	certain sources or conditions in an equipment, or performs a transfer operation automatically.
84	**operating mechanism** is the complete electrical mechanism or servo-mechanism, including the operating motor, solenoids, position switches, etc., for a tap changer, induction regulator or any piece of apparatus which has no device function number.
85	**carrier, or pilot-wire, receiver relay** is a device which is operated or restrained by a signal used in connection with carrier-current or d-c pilot-wire fault directional relaying.
86	**locking-out relay** is an electrically operated hand or electrically reset device which functions to shut down and hold an equipment out of service on the occurrence of abnormal conditions.
87	**differential protective relay** is a protective device which functions on a percentage or phase angle or other quantitative difference of two currents or of some other electrical quantities.
88	**auxiliary motor, or motor generator** is one used for operating auxiliary equipment such as pumps, blowers, exciters, rotating magnetic amplifiers, etc.
89	**line switch** is used as a disconnecting or isolating switch in an a-c or d-c power circuit, when this device is electrically operated or has electrical accessories, such as an auxiliary switch, magnetic lock, etc.
90	**regulating device** functions to regulate a quantity, or quantities, such as voltage, current, power, speed, frequency, temperature, and load, at a certain value or between certain limits for machines, tie lines or other apparatus.
91	**voltage directional relay** is a device which operates when the voltage across an open circuit breaker or contactor exceeds a given value in a given direction.
92	**voltage and power directional relay** is a device which permits or causes the connection of two circuits when the voltage difference between them exceeds a given value in a predetermined direction and causes these

/more/

Table 10-1. Protective Device Numbering and Functions
(concluded)

DEVICE NUMBER	DEFINITION AND FUNCTION
92 (con't.)	two circuits to be disconnected from each other when the power flowing between them exceeds a given value in the opposite direction.
93	field changing contactor functions to increase or decrease in one step the value of field excitation on a machine.
94	tripping, or trip-free, relay is a device which functions to trip a circuit breaker, contactor, or equipment, or to permit immediate tripping by other devices; or to prevent immediate reclosure of a circuit interrupter, in case it should open automatically even though its closing circuit is maintained closed.
95 **96** **97** **98** **99**	Used only for specific applications on individual installations where none of the assigned numbered functions from 1 to 94 is suitable.

notes: [1] A similar series of numbers, starting with 201 instead of 1, shall be used for those device functions in a machine, feeder or other equipment when these are controlled directly from the supervisory system. Typical examples of such device functions are 201, 205, and 294.

[2] A suffix X, Y, or Z denotes an auxiliary relay.

[3] TC refers to trip coil.

[4] CS refers to control switch.

[5] N, G refers to neutral and ground respectively.

Overcurrent relays are available with inverse, very inverse and extremely inverse time current characteristics. The very inverse time current characteristic is the frequent choice when detailed system information is not available. The very inverse characteristic is most likely to provide optimum circuit protection and selectivity with other system protection devices. When

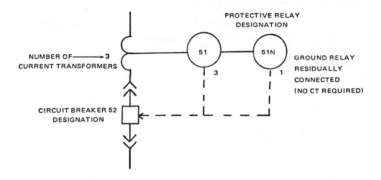

Figure 10-1. One-Line Diagram Showing Protective Relays

coordinating with fuses it may be necessary to use the extremely inverse characteristic.

Figure 10-2 illustrates the characteristics of electromechanical overcurrent protective relays. The overcurrent relays are current sensitive and require a seal-in contact to keep them energized after activation.

A common way in which power is supplied to a breaker for tripping purposes is through a d-c source using 125- or 250-volt station battery.

Figure 10-3 shows a typical electromechanical overcurrent relay schematic. Relay 51X is used as the seal-in relay which holds in the circuit until it is reset. For a three-phase circuit three of these relays are required.

Figure 10-4 illustrates a typical tripping circuit. Notice that protective relay contacts are connected in parallel so that any one will trip the breaker under fault conditions. The control switch (CS) can be manually used to trip the breaker. In order to simplify the protective relay elementary, many times a single contact is used to represent the three. Figure 10-5 presents a simplified protective relay elementary. To close the breaker, frequently just a manual close switch is used. A typical elementary used to close the circuit breaker is illustrated in Figure 10-6.

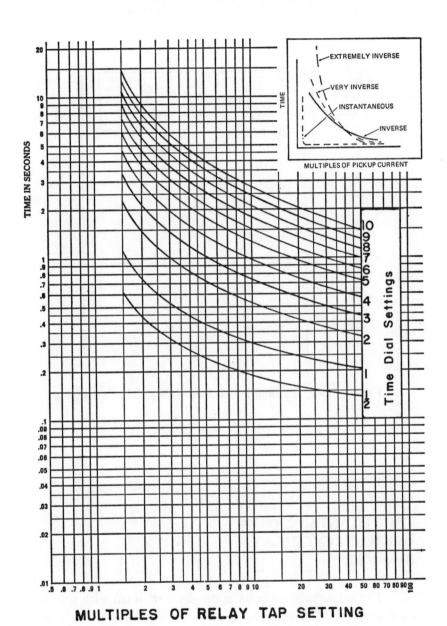

MULTIPLES OF RELAY TAP SETTING

Figure 10-2. Characteristics of Overcurrent Relays

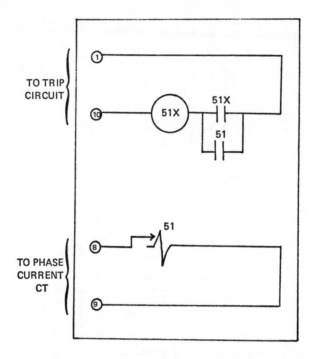

Figure 10-3. Typical Protective Relay (51) Schematic

THE INSTANTANEOUS
OVERCURRENT RELAY
(Device 50)

The instantaneous relay is usually combined with the over-current relay. The instantaneous attachment can be used or it can be disconnected from service if it is not required. Figure 10-7 illustrates a typical one-line diagram and schematic utilizing the 50, 51 and 51N protective relays.

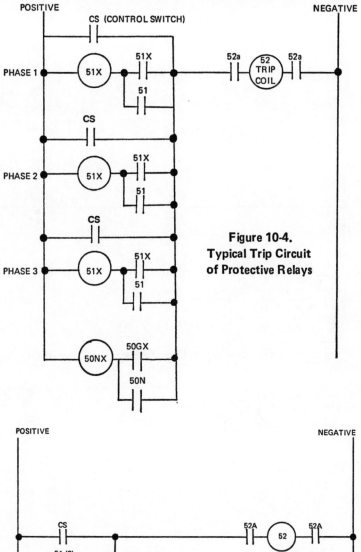

Figure 10-4.
Typical Trip Circuit
of Protective Relays

Figure 10-5.
Simplified Protective Relay
Elementary

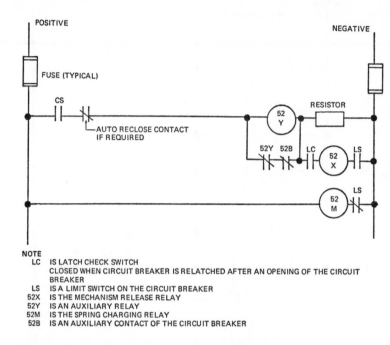

NOTE
LC   IS LATCH CHECK SWITCH
     CLOSED WHEN CIRCUIT BREAKER IS RELATCHED AFTER AN OPENING OF THE CIRCUIT
     BREAKER
LS   IS A LIMIT SWITCH ON THE CIRCUIT BREAKER
52X IS THE MECHANISM RELEASE RELAY
52Y IS AN AUXILIARY RELAY
52M IS THE SPRING CHARGING RELAY
52B IS AN AUXILIARY CONTACT OF THE CIRCUIT BREAKER

Figure 10-6. Typical Close Elementary of a Circuit Breaker

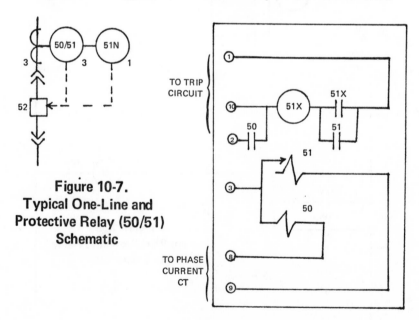

Figure 10-7.
Typical One-Line and
Protective Relay (50/51)
Schematic

THE GROUND OVERCURRENT RELAY
(Devices 50N and 50G)

Ground-fault protection has been required since the 1971 NEC. Ground-fault protection saves lives by minimizing damage to circuit conductors and other equipment safeguarding persons who may simultaneously make contact with electrical equipment and a low resistance path to ground. Ground overcurrent protection can be provided either by overcurrent or instantaneous overcurrent relays. There are three common connections used for ground overcurrent relays; namely, residual connection, ground sensor connection, and the neutral CT connection.

Figure 10-8 illustrates the residual grounding scheme. Notice that the one-line diagram of Figure 10-7 has been detailed to show the three-phase connections. The residual relaying scheme detects ground-fault current by measuring the current remaining in the secondary of the three-phase of the circuit as transformed by the current transformers. Care must be taken to set the pick-up of the relay above the level anticipated by unbalanced single-phase loads. Due to the possible unbalances caused by unequal current transformer saturation on phase faults and transformer energizing inrush currents, the instantaneous overcurrent relay is seldom used.

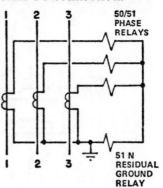

Figure 10-8.
Residual Grounding Scheme

The ground sensing scheme is illustrated in Figure 10-9. This scheme uses zero sequence current transformers to detect on ground faults the unbalances in the magnetic flux surrounding the three-

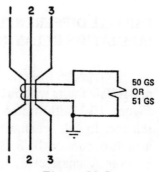

Figure 10-9.
Ground Sensor Scheme

phase conductors. Zero sequence current transformers detect when the vectorial summation of the currents is not zero.

The instantaneous or overcurrent relay can be used with this scheme. The installation of the zero-sequence window current transformer should not enclose the equipment ground conductor or the conductor shielding. With the ground sensing scheme it is possible to detect and clear system faults as small as 15 amperes.

The neutral grounding scheme illustrated in Figure 10-10 is used commonly with resistively grounded transformers. In this scheme the ground-fault current is sensed by the current transformer in the resistively grounded neutral conductor.

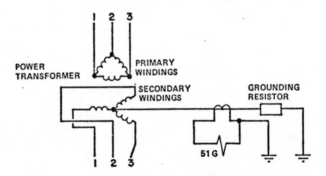

Figure 10-10. Neutral Grounding Relay Scheme

PARTIAL DIFFERENTIAL OR SUMMATION RELAYING

This protective relaying scheme is commonly used to detect and isolate faults without affecting other portions of the system. Figure 10-11 illustrates a typical partial differential relaying scheme. In this scheme the tie breakers are nominally closed. If a fault occurs on BUS G, breakers C and D should trip leaving breaker A unaffected. Likewise a fault on BUS F should trip breakers A and C and leave breaker D unaffected. One way of accomplishing this is to connect the current transformers of

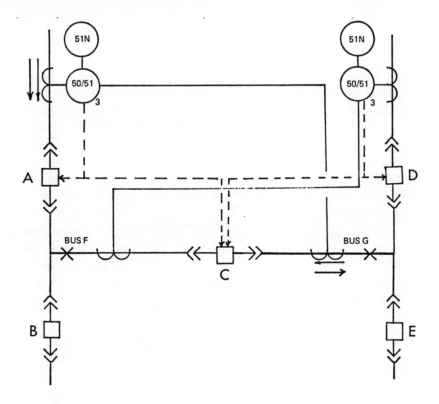

Figure 10-11. Partial Differential Protective Relay Scheme

protective relays such that they will only pick up when the fault currents through the pair of current transformers flow in opposite directions. For example, a fault on BUS F will cause fault currents flowing *to* the fault. The current transformers associated with protective relays for breaker A will sense currents in opposite directions thus activating these relays. On the other hand, a fault on BUS G will cause currents to flow in the same direction through the current transformers, thus these relays will not operate under this condition.

DIFFERENTIAL PROTECTIVE RELAY
(Device 87)

The differential protective relay is used for protecting a-c rotating machinery, generators and transformers. This relay operates on the difference between two currents. A typical application is illustrated in Figure 10-12. This figure shows the differential principle applied to a single-phase winding of electrical equipment such as a generator. In this application a current balance relay is used to provide what is called "percentage differential" relaying. The current transformers are connected to the equipment to be protected.

The current from each transformer flows through a restraining coil. The purpose of the restraining coil is to prevent undesired relay operation as a result of a mismatch in current transformers. When a fault does occur the operating relay sees a percentage increase in current and the relay operates.

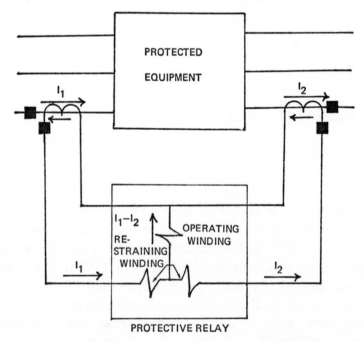

Figure 10-12. Differential Protective Relay Scheme

Notice on Figure 10-12 the introduction of polarity marks on the current transformers. Polarity identification marks are as follows:

• Current flows into the polarity mark for primary connections. Current flows out of the polarity mark for secondary connections.

• Voltage drop from polarity to nonpolarity for primary and secondary connections.

See Figure 10-13 for an illustration of polarity marks.

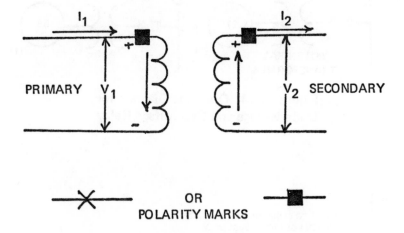

Figure 10-13. Polarity and Circuit Diagram

UNDERVOLTAGE RELAY (Device 27)
AND OVERVOLTAGE RELAY (Device 59)

The undervoltage and overvoltage relays are used wherever protection is required for these conditions. These relays usually operate continuously energized and are adjusted to drop out at any voltage within their calibration range. Figure 10-14 illustrates a one-line diagram for undervoltage and overvoltage relays. In this particular scheme, breaker A is tripped by undervoltage relay. An auxiliary contact from breaker A trips breaker B after breaker A is tripped. The overvoltage relay closes breaker B when the preset voltage level is reached. The characteristics of undervoltage relays are indicated in Figure 10-15.

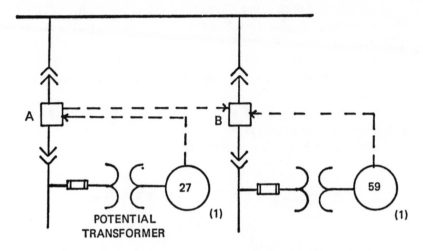

**Figure 10-14. One-Line Diagram for
Undervoltage and Overvoltage Relay**

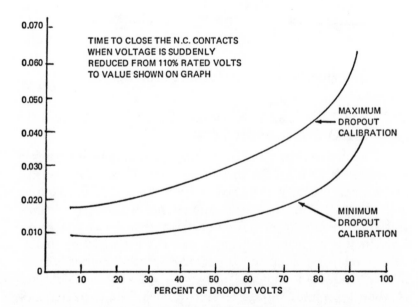

Figure 10-15. Characteristics of Undervoltage Relays

APPLYING SOLID-STATE PROTECTIVE RELAYS

Solid-state protective relays offer significantly improved characteristics over electromechanical relays and are available for the most important areas of the system. Solid-state techniques allow improvement in sensitivity and temperature stability, plus effective transient surge protection.

The main advantages in using solid-state relays are:
- *Flexible settings*
- *Dynamic performance*
- *Improved instantaneous*
- *Very low burden*
- *Easy testing*
- *Reduction in panel space*
- *Improved indication*
- *Better immunity to shock*
- *Good repeatability*

These relays are available in time overcurrent relays, instantaneous overcurrent relays, voltage relays, directional relays, reclosing relays, ground-fault relays, direct-current relays and timing relays.

Some of the advantages of solid-state relays are summarized below.

EASY TESTING
AND FLEXIBLE SETTINGS

Installation testing is performed by depressing test buttons on the relay; thus test equipment is not required. These test buttons allow for initial settings, operational and wiring checks, and maintenance tests. A rough timing check can be performed with the second hand of an ordinary wrist watch. More precise tests can also be made.

DYNAMIC PERFORMANCE

The solid-state relay can be thought of as a fine-tuned electromechanical protective device. The conventional electromechanical overcurrent relay is the induction disc type. The induction disc element which is either copper or aluminum rotates between the pole faces of an electromagnet. There are two methods commonly used to rotate the induction disc. The shaded pole method is illustrated in Figure 10-16. In this method a portion of the electromagnet pole face is short-circuited by a copper ring or coil to cause the flux in the shaded portion to lag the flux in the unshaded portion.

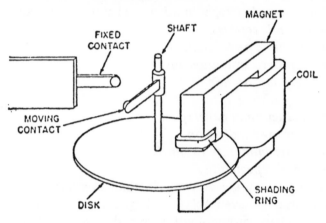

Figure 10-16. Shaded Pole Induction Disc Type Overcurrent Relay

The second method is referred to as the wattmetric type and uses one set of coils above and below the disc. In both methods the moving contact is carried on the rotating shaft. In the induction-disc type of overcurrent relay, the disc continues to rotate after the starting current has decreased to a low value. This overtravel means that in order to avoid nuisance tripping the time dial setting of the relay is set several positions higher than is desirable. Solid-state relays do not have rotating parts, thus the problem of overtravel does not occur. This means that the characteristics of the relay can be adjusted to what is required without allowances for dynamic effects such as overtravel.

The solid-state overcurrent relay consists of printed circuit boards that produce a d-c output voltage when the input a-c current exceeds a given value. Each overcurrent function usually consists of an input transformer, overcurrent module, and a resistor-zener-diode protective network. The overcurrent module consists of a setting circuit, phase-splitter circuit, sensing circuit, amplifier circuit, feedback circuit and an output circuit. The overcurrent module can either be a single-input module with one output or a dual-input module with a single output.

LOW BURDEN

Low-volt ampere (burden) requirements for protective relays means that more relays may be connected in series, the lowest tap of a multi-ratio CT can be used, and the auxiliary CT can be stepped up for residual current sensitivity. This translates into dollar and space savings as a result of low-volt ampere requirements on instrument transformers. Another savings is that bush-mounted current transformers are not required with solid-state relays. Solid-state relays are compatible with bushing-mounted current transformers in low- and medium-voltage switchgear applications.

REDUCTION IN PANEL SPACE

A comparison of solid-state relay sizes with their electro-mechanical counterparts indicates a space savings of at least one-third. For complex relay schemes space savings can be 75 per cent or more.

IMPROVED INDICATION

The solid-state protective relay has a target which operates independently of the trip-coil current and depends only on the proper functioning of the relay. This helps in troubleshooting in the event of a broken trip-coil connection. With the new design it is possible to have an indicator without a seal-in contact in parallel with the relay's measuring contacts. Often this results in simplification of complex schemes.

BETTER IMMUNITY TO SHOCK

Since the solid-state relay has no moving parts it is in many cases better suited for earthquake-prone locations. Solid-state relays have been tested and have withstood accelerations of up to 10g and higher.

RELIABILITY

Solid-state devices are reliable, have good repeatability, and are economical for industrial applications. The major protective relay manufacturers now offer solid-state relays as part of their line. Over the last few years solid-state protective devices have proven to be a new tool for system protection.

11

Reducing Electrical Utility Costs

With the advent of an energy consciousness in the United States, it has become apparent that electrical energy costs must be managed. As electrical utility rates increase, there is an increasing pressure to conserve energy and to reduce electrical costs wherever possible. This chapter will illustrate several ways to reduce electrical utility costs.

ANALYZING UTILITY RATE STRUCTURES

It is important to understand how the plant is billed before initiating an energy conservation program. Several items which should be determined are:

• Does the contract contain a *power factor correction clause?* Most utility companies have a penalty clause if the power factor is less than .85.

• What is the *demand charge?* The demand charge is based on the maximum kilowatt requirement over a 15-, 30- or 60-minute interval.

• What is the *load factor charge?* Load factor is the ratio of the average load over a designated period to the peak demand load occurring in that period.

- What is the energy *usage rate?*
- Is there a penalty clause based upon *time of day?*
- Does the utility contract have a *"ratchet" rate?* A "ratchet" rate means that the demand rate is based on the maximum demand for any period. It cannot be lowered for an entire year and only then if the new demand is lower.

Both the time of day clause and ratchet rate are relatively new concepts utilities are experimenting with.

UTILIZING ENERGY EFFICIENT PRODUCTS

Wherever possible consideration should be given to increasing the efficiency and/or the power factor of equipment specified. Approximately 50 percent of the industrial plant energy consumption is due to electric motor drives. Thus, it is important to evaluate the energy efficiency of the product specified. A motor's efficiency rating indicates how well it converts electrical energy into mechanical energy. As indicated in Table 3-1, Chapter 3, motors below 30 HP have a relatively poor efficiency and power factor. Motors operating at partial load cause the power factor to be even worse. For years motor designers have let first-cost considerations determine design.

MOTOR EFFICIENCY IMPROVEMENT

To improve the motor efficiency internal losses must be minimized. Core losses occur because of hysterisis effects and eddy currents. Eddy current losses are reduced by using thinner gauge laminations and steel with improved core-loss properties. Another way to reduce losses is to increase the magnetic core which in turn increases the area and reduces the flux density. To reduce the losses generated as a result of line current passing through the stator winding, the resistance of material is reduced. This can be accomplished either by using a higher conductivity material or by increasing the cross-sectional area of the conductor. Another way is to reduce the magnetizing portion of the current by shortening the air gap or by reducing flux density.

MOTOR POWER FACTOR IMPROVEMENT

The best way to achieve power factor improvement without affecting motor efficiency adversely is to add material. Adding material by increasing the length of the stator and rotor cores increases both the power factor and efficiency.

USE OF ENERGY EFFICIENT MOTORS

Energy efficient motors are available from manufacturers such as Gould, Inc. Energy efficient motors are approximately 30 percent more expensive than their standard counterpart. Based on the energy cost it can be determined if the added investment is justified. With the emphasis on energy conservation new lines of energy efficient motors are being introduced. Figures 11-1 and 11-2 illustrate a typical comparison between energy efficient and standard motors.

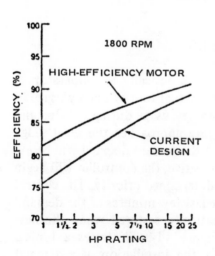 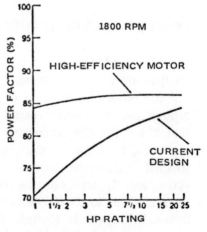

Figure 11-1.
Efficiency vs Horsepower
Rating (Dripproof Motors)

Figure 11-2.
Power Factor vs Horsepower
Rating (Dripproof Motors)

SYSTEM POWER FACTOR IMPROVEMENT

As illustrated in Chapter 3, another way to improve the plant power factor is by using capacitors. Capacitors supply leading Kvars to the system. Another method to improve the plant power factor is to use synchronous motors.

REDUCING PEAK DEMAND

One of the biggest energy cost-saving potentials is to reduce peak demands. The simplest method to do this relies upon manually scheduling activities so that big power users do not operate at the same time. This is sometimes possible during initial plant start-up where one system can be operated when another is down. The second method relies upon automatic controls which shut off nonessential users during peak periods. Nonessential users such as heating, ventilation and air-conditioning equipment can be automatically controlled through packaged equipment such as load-demand controllers.

LOAD DEMAND CONTROLLERS

The load demand controller is basically a comparator. A comparison is made between the actual rate of energy usage to a predetermined ideal rate of energy usage during the demand interval. As the actual usage rate approaches the ideal usage rate, the controller determines if the present demand will be exceeded. If the determination is positive, the controller will begin to shed loads based upon a predetermined priority. The control action usually occurs during the last few minutes of the demand interval. The loads are automatically restored when the new demand interval is started. Figure 11-3 illustrates a typical demand chart before and after the installation of a demand controller.

There are several types of demand controller on the market. Careful consideration should be given to the type of controller

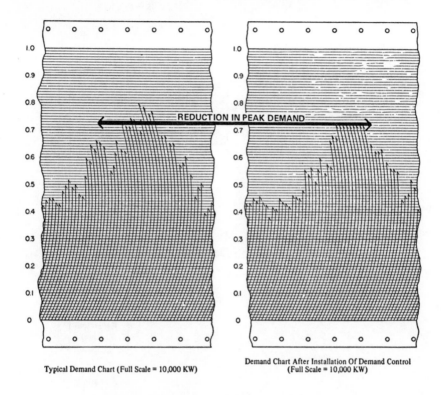

Typical Demand Chart (Full Scale = 10,000 KW)

Demand Chart After Installation Of Demand Control
(Full Scale = 10,000 KW)

**Figure 11-3. Typical Demand Chart Compared
After Installation of a Demand Control**

specified. Cost and proper application are the primary criteria for choosing the demand controller for the plant. Several commonly used types are:

- *Packaged solid-state controllers*
- *Time-interval reset type*
- *Sliding-window type*
- *Computer*

The *time-interval reset type* controller needs two input signals as illustrated in Figure 11-4. The inputs represent kilowatt-hour energy usage and an interval signal pulse. Before purchasing this unit, it should be verified that the utility company will permit use of the synchronous pulse with the demand con-

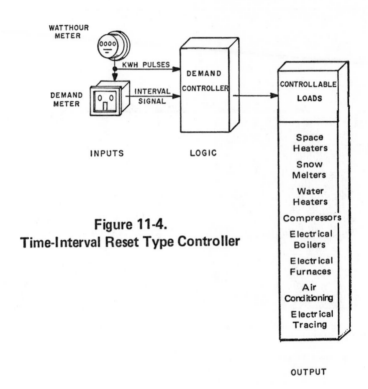

Figure 11-4.
Time-Interval Reset Type Controller

troller. Several utilities do not permit this signal to be used; thus another type of controller may be required.

The *sliding-window type* does not require a synchronous pulse. In the sliding-window type power consumption is monitored continuously. The width of the sliding window can be set from five to sixty minutes in five-minute increments. The energy consumed during this interval is proportional to demand. Loads are shed and restored based on predetermined upper and lower thresholds. Predicted usage is calculated based on the consumption during the most recent quarter window width. A scheduling clock can be built into these units allowing for turning loads on and off during specific times of the day.

Computers used for load shedding can also serve several other functions. Computer systems presently offered can be used in conjunction with security and fire-alarm protection. These units also allow for better reporting of energy consumption through audit reports. One feature of computer load shedding is the floating target adjustment. This allows the system to automatically respond to changing conditions in the facility such as weekends, holidays, and unexpected work stoppages.

LIGHTING EFFICIENCY

The basics of lighting were reviewed in Chapter 5. To reduce the operating costs of lighting, the following steps should be considered:

- Use appropriate task lighting levels for work areas and reduce lighting in aisles to one-third of the task lighting level.
- Choose a lamp based on the life-cycle cost. High intensity discharge lamps offer a high lumen output and long life.
- Choose a luminaire with the best coefficient of utilization for the working space.
- Initiate group relamping and cleaning programs so that the number of luminaires required in the initial design can be reduced.
- Choose a luminaire which resists dirt build-up.
- Provide adequate switching and/or automatic controls so that when lighting is not required it can be easily turned off.

To achieve an efficient lighting design the engineer must question "rules of thumb" and previous lighting practices, and determine the best system for the particular operation. In addition, the psychological effect on workers and the system reliability should be considered prior to making a change. Marginal energy savings may result from reducing lighting levels in areas which are used only a few hours a day. On the other hand, reduced lighting at some tasks may result in poor quality work, thus negating any savings. This is usually not an effective way to reduce energy consumption. A far better approach is to replace the lamps with a higher lumen lamp with reduced wattage.

Where lamps have been removed the fixture ballast should be disconnected. If the ballast is not disconnected, the ballast will continue to draw power, thus negating a portion of the savings. An energized ballast can also affect the power factor adversely.

A product known as the "Phantom Tube" has been developed by Developmental Sciences, Inc., CA. If a fluorescent tube needs to be removed the phantom tube can be used in its place, thus saving energy, not adversely affecting power factor and maintaining uniform reduced lighting.

ANALYZING ENERGY USAGE CHARTS

A good energy detective's tool is the electrical energy usage charts. From these charts can be determined wastes such as equipment running when not needed on holidays, weekends and at night. In cases where load demand controllers are used to reduce peak demand by turning off heating, ventilation and air conditioning units substantial savings will also be made by reducing energy usage.

SUMMARY

Energy consumption in lighting systems should be reduced even further with a number of new developments. A new high efficient phosphor for fluorescent lamps has been developed. As an example of the potential savings a 35 watt fluorescent lamp coated with phosphor can produce 97 percent as much light initially as a standard 40 watt fluorescent lamp. General Electric estimates that if all of the 800 million 40 watt lamps now installed in commercial buildings and industrial plants in the United States were switched to the new lamp 22 billion Kwh of electricity per year would be saved.

A new electrodeless and ballastless fluorescent has been developed which fits into a standard incandescent fixture. The bulb has no incandescent filament; thus fluorescent lighting efficiency will be available from a bulb that looks like and fits Standard incandescent sockets. The new fluorescent bulb uses a small VMOS electronic package in the bulb's base to produce an RF signal that sets up the magnetic field which excites surrounding atoms in fluorescence.

With many new energy efficient devices being introduced care should be taken, so as to avoid using devices which do not live up to their so-called claims. Good engineering judgement will usually uncover unsubstantiated energy conservation products. The key to energy conservation is not to live in darkness. Rather it is to apply the systems more efficiently so that the quality of the installation is maintained while conserving energy.

References

1. Burrows, W. W., OSHA: "Recognize and Correct Electrical Violations," *Maintenance Engineering,* 1973.
2. Burrows, W. W., OSHA: "Install and Maintain Valid Electrical Grounds," *Maintenance Engineering,* 1973.
3. Burrows, W. W., OSHA: "Checking Out Your Overcurrent Devices and Transformers," *Maintenance Engineering,* 1973.
4. Burrows, W. W., "Bring Your Switches, Boxes and Outlets into OSHA Compliance," *Maintenance Engineering,* 1973.
5. Scarnecchia, V., OSHA: "Inspect, Check and Correct Wiring Violations," *Maintenance Engineering,* 1973.
6. Magison, E. C., **Electrical Instruments in Hazardous Locations,** Instrument Society of America, Pittsburg, Pa., 1974.
7. **Recommended Practice for Classification of Areas for Electrical Installations in Petroleum Refineries,** API RP 500A, American Petroleum Institute, 1966.
8. Redding, P. J., *Intrinsic Safety − The Safe Use of Electronics in Hazardous Areas,* McGraw Hill, London, 1971.
9. **National Electrical Code,** 1971-75-78 Editions, National Fire Protection Association, Boston, Mass.
10. **Occupational, Safety and Health Standards** (29 CFR 1910), U. S. Department of Labor, January 1976.
11. **Electrical Power Distribution for Industrial Plants,** Institute of Electrical and Electronic Engineers, New York, N. Y., 1964.
12. **Illuminating Engineering Society Lighting Handbook,** Illuminating Engineering Society, New York, N. Y., 1972.
13. Fitzgerald, A. E. and Kingsley, C., *Electric Machinery,* McGraw-Hill, New York, N. Y., 1952.

14. Brenner, E. and Javid, M., *Analysis of Electric Circuits,* McGraw-Hill, New York, N. Y., 1959.
15. Beeman, D., *Industrial Power Systems Handbook,* McGraw-Hill, New York, N. Y., 1955.
16. Industrial Systems Data Book, General Electric Company, New York, N. Y.
17. Croft, T.; Carr, C.; Watt, J., *American Electrician's Handbook,* McGraw-Hill, New York, N. Y., 1970.
18. Fink, D. G. and Carrol, Jim, *Standard Handbook for Electrical Engineers,* McGraw-Hill, New York, N. Y., 1969.
19. Abbott, A. L. and Stetka, F., *National Electrical Code Handbook,* McGraw-Hill, New York, N. Y., 1964.
20. Westinghouse Lighting Handbook, Westinghouse Electric Co., Bloomfield, N. J.
21. Energy Conservation Program Guide for Industry and Commerce, NBS Handbook 115, U. S. Department of Commerce.
22. Roe, L. B., *Practices and Procedures of Industrial Electrical Design,* McGraw-Hill, New York, N. Y., 1972.
23. Steel Electrical Raceways Design Manual, American Iron and Steel Institute, New York, N. Y., 1968.
24. Thumann, A., "Logic Language Aids Consultants," *Electrical Consultant,* March 1973.
25. Thumann, A., "For Greater Control Sequence Flexibility, Use the Programmable Controller," *Electrical Consultant,* Atlanta, Ga., January 1974.
26. Thumann, A., Instrumentation—Growing Field for Electrical Consultant, *Electrical Consultant,* August 1974.
27. Applied Protective Relaying, Westinghouse Electric Corporation, 1964.
28. James, Waldrun E., "Solid State Protective Relaying," *Electrical Consultant,* June 1976.

Index

Analyzing Electrical Control Circuits 169
Analyzing Logic Diagrams 181
Application Table for Short Circuit Current
     & Breakers 100, 101, 102

Bid Analysis
     Analysis 124
     Forms 126
Breaker Advantages 117
Breakers for Motors 118
Breakers for Motor Control Centers 116
Breakers for Substations 99

Conductor
     Derating Due to Ambient 156
     Derating Due to Number 156
     Sizing 157
     Use in Raceways 152

Conduit
     Power and Control 163
     Sizing 158

Control
     Devices 168, 169
     Symbols 170

Delta Connected Transformer 90
Demand Charge 239

Demand Controllers 242, 243
Design Activity Manhours 75
Design Organizations 79
Drawing List 87

Electrical Rooms 116
Elementaries
    Elementary Diagram 69
    Interlocking 172
    Reversing Motors 172
    Two Speed Motors 173
Energy Conservation
    Analyzing Utility Rate Structures 239
    Demand Controller 242, 243
    Lighting 151, 245, 246
    Motor Efficiency Improvement 240
    Motor Power Improvement 241
Engineering Activities 67, 78

Full Voltage Revising Elementary 173
Full Voltage Non-Reversing Elementary 171
Fuse Data For Motors 119
Fuses Advantages 117

Grounding Drawing 69

Hazardous Areas
    Class Hazard Designations 205
    Classification of Areas 207
    Division Classification 206
    Installation 206
    Group Hazard Designations 205
Holding Coil 171

Instrumentation
    Control Loop 197, 199
    Flow Meter 200
    Magmeter Type 200

Orifice Meter 201
Pressure Control 201
Temperature Control 201
Intrinsic Safety
    Benefits 211
    Installation 210
    Application 208

Lighting
    Choosing a Lighting System 131
    Choosing Levels 130
    Circuitry 146
    Coefficient of Utilization 135
    Drawings 69, 148
    Fixture Layout 144
    Footcandle 130
    Lighting Methods 135
    Lighting Panel 74
    Light Loss Factor 143
    Lighting Efficiency 245
    Lumen 130
Load Factor Charge 239
Logic Diagram
    Analysis 177, 181
    Compressor Logic 182
    Description of Operation 174
    Symbols 178
    Use with Mini Computers 176

Motor
    D.C. Motors 97
    Efficiency 240
    Horsepower 91
    Power Factor Improvement 241
    Squirrel Cage Induction Motor 97
    Synchronous Motors 97
    Two Speed 173
    Voltages 98

Motor Control Centers
Layout 121, 122
Oneline 123

One-Line Diagrams
How to Draw 109
Typical One-Line Diagram 68
Organization 79, 80
OSHA
Bring Switches, Boxes and Outlets into
Compliance 38
Checking out Overcurrent Devices and
Transformer 25
Check List of NEC 7
Commonly Cited Electrical Violations 3
Degree of Compliance 4
Install and Maintain Valid Electrical Grounds 12
OSHA: Inspect, Check and Correct Wiring
Violations 57
Provisions for Transformer Vaults 36

Power Factor
Correction Clause 239
Correction 92
Importance 92
Power Flow Concept 96
Power Layouts 69, 165
Power Triangle 88
Primary Selective System 110
Programmable Controller 192
Protective Relays
Characteristics 225
Device Numbers 213-223
Differential Protective Relay 232
Ground Overcurrent Relay 229
Ground Sensor Scheme 229
Instantaneous Overcurrent Relay 226
Overvoltage Relay 233

Partial Differential Relaying 230
Polarity Marks 233
Solid State 235
Undervoltage Relay 233

Scheduling
  Critical Path............................. 80, 81, 84
Short Circuit Currents
  Faults Occurrence 99
Simple Radial System 110
  Secondary Selective System 111
Substation
  Ratings 99
Switchgear Breakers 96

Transformer
  Ratings 100
  Windings 89

Voltage Considerations 109, 110

Wiring Methods 203
Wye-Connected Transformer 90

Zonal Cavity Method 135